Researching Tourism, Leisure and Hospitality for Dissertations and Theses

Second Edition

Peter Mason

(G) Goodfellow Publishers Ltd

(G) Published by Goodfellow Publishers Limited,
26 Home Close, Wolvercote, Oxford OX2 8PS
http://www.goodfellowpublishers.com

British Library Cataloguing in Publication Data: a catalogue record for this title is available from the British Library.

Library of Congress Catalog Card Number: on file.

ISBN: 978-1-915097-00-2

DOI: 10.23912/9781915097002-4976

Design and typesetting by P.K. McBride, www.macbride.org.uk

Cover design by Cylinder

Printed by Lightning Source

Contents

List of Case Studies

List of figures

List of tables

Preface to the Second Edition

This book is based on teaching, supervising and examining undergradute and postgraduate students for over thirty five years in the UK, France, Australia and New Zealand. It is aimed primarily at undergraduate and postgraduate students in English speaking countries, or where English is the language of the classroom. However, it should also be useful as an introduction to those students embarking on a PhD in the related fields of Tourism, Leisure or Hospitality.

The key areas that the book covers are:

- Becoming self-aware of your research capability
- Developing your formal, academic writing style
- Selecting and refining your research topic
- Creating your Literature Review
- Writing your Methodology
- Presenting your Results
- Analysing and Discussing your Results
- Presenting Conclusions

The book's structure mirrors closely the chapters of most student research projects, and each text chapter includes a section on how to write the corresponding part in the student's project This approach follows students' comments when I have taught them research methods and research project preparation.

I have used several of my own research projects as case studies, and drafts of some students' work, to exemplify certain research processes and techniques, as well as to provide opportunities to consider the advantages and disadvantages of the particular approaches.

This second edition has been completely revised and updated. The major changes are the addition of more material on on-line surveys and new case studies, but in particular a new chapter (Chapter 6) on the design and conduct of mixed methods research and a new Chapter 10 which discusses the analysis process in mixed methods research.

The book is written in, what is intended to be, a clear, straightforward, easy to comprehend style and I hope it offers useful advice and assistance that enables students to produce worthwhile and successful research projects.

Acknowledgements

This book did not come into being without the assistance of a number of people.

I would like to give a special thanks to my editor Sally North, who has not only helped with this second edition, but was there for the first edition and offered great support in the editing of several other books I have written. I would also like to thank the man whose job it is to turn my far from high quality word processing into an attractive looking book – thanks Mac!

The book is based largely on my teaching and supervising students, which has taken place in several locations, and I would like to thank specfically undergraduate and postgraduate students at Bedfordshire University, London Metropolitan University, London South Bank University and Westminster University in the UK and those in the Business School, Massey University, New Zealand and the University of Rijeka Faculty of Tourism and Hospitality, Croatia. I would also like to thank PhD students I have either supervised or examined at Bedfordshire University, Bournemouth University, University of Central Lancashire, King's College, London, Massey University, New Zealand and Monash University, Australia.

I have made detailed reference to the work of three individual students (draft questionnaire survey, draft interview schedule and PhD mixed methods research) but for ethical reasons these students remain anonymous.

There were several academics I thanked in relation to the first edition of his book (Andrew Holden, Robert Maitland, Richard Sharpley, Duncan Tyler and I-Ling Kuo) and I would like to repeat my thanks. I would also like to give my thanks to the anonymous academic reviewers of my proposal for this second edition.

I would like to thank my wife Patsy who noticed, but did not always comment on my absence at the computer, my son Will who is now an academic and my daughter Jess, a musician, who also works in the 'events industry'. This book is dedicated to my first grandchild, Hector!

Whatever blemishes may remain in his book, however, are ultimately my responsibility!

Peter Mason, Catfield, Norfolk February, 2022.

1 An Introduction to Research and Your Study

Introduction

Almost everywhere in the English-speaking world, and in many other parts as well, it is now the case that as a university student of tourism (or related subjects, such as hospitality or leisure) you are required to produce an individual research project. At whatever level, be it undergraduate or post-graduate (Masters), this is likely to be a daunting task for many of you. It is unlike any assignment that you have had to produce before. It is much longer, requires a good deal of originality, and will almost certainly involve you in the gathering of primary data and hence the use of research methods, which you may know little about and believe that you have never used before!

There are a number of terms for the individual student piece of research: 'dissertation', 'thesis', 'report' or as stated above 'project'. Whatever the term used, it is a substantial piece of independent research with the student leading and directing it (Coles et al., 2013). As an undergraduate, you will usually undertake a dissertation in the final year of study. A dissertation for a Masters student will normally be done as the final piece of work, but spread over several weeks or months. A thesis is the name given, in many countries, to the major independent study undertaken by a PhD student. In this book, the terms 'individual research project' or 'individual study' (frequently summarised as either 'study' or 'project') are used for a student's individual independent piece of assessed research. However, the book is aimed primarily at those at undergraduate or postgraduate level who are writing a dissertation.

For your individual research project, you will be required to read more extensively than for other written assignments. The reading will not be confined to one particular, very narrow theme or topic, as may be the case in relation to an individual assignment but will be concerned with the wider literature that is relevant to the chosen topic. In addition to the literature on your topic or issue, there will also be the need to read literature on the methodology and techniques to be used to conduct your research, as well as literature on how to analyse the results of this research.

To reflect the importance of the research project, it frequently has double weighting in terms of overall assessment marks, is often represented as a double module, and may be worth up to one third of the marks for a course. Therefore, the research project may often be equivalent to at least two, but possibly four 'normal' taught modules and is usually to be completed in the third semester of the overall course programme. The project is the only part of the course concerned solely with research. Additionally, you are left, to a very great extent, to your own devices and will be required to work largely alone over several months with only occasional input from a supervisor.

To many students, the structure of the research project will make it seem as if there are several assignments together, not just one, as each chapter will appear to require at least as much time and effort as an individual assignment. In general, you will be familiar with a fairly short time frame of just a few weeks to prepare for, write and submit a 'standard' assignment. However, your research project will often take months to prepare, write and present.

Whilst in many courses, which are usually made up of related modules, it is unusual to have prerequisite modules. However, in relation to the research project there is almost always a pre-requisite – this is the 'Research Methods' module. It is often the case that one or more of your taught modules has given you the initial idea for your chosen study topic. One of the reasons that the research project is likely to be the last aspect of your course, is that you will usually need time to reflect on your taught modules and your reading, to come up with your topic. However, you may find that you are working on your research project, while still being taught other modules on your course.

Also, in relation to the research project, students are given a supervisor and usually have one-to-one tutorials. For many students this is not just a new experience, but one accompanied by much anxiety. A student will usually be required to prepare some written work, send this to their supervisor in advance and then be prepared to answer questions about what they have written, without knowing in advance what those questions will be.

The way in which the research project is assessed is different from other assignments in terms of the way it is marked and who is involved. You may be asked to prepare individual sections or chapters of the study, and then these drafts are 'marked' or commented on by the supervisor, before being discussed individually with you. Assignments linked to taught modules are usually marked by the person who lectures on the module. Under normal circumstances, the research study will be marked by the supervisor, and also at least one other person, based at the same institution, who has not been involved in assisting you in writing the project. It is also quite likely that an external examiner will be involved. This person is external to the university or college where you are studying and is involved, not just to consider marks of at least a sample of the research project, but to ensure consistency of standards across all universities in that region or country.

Increasingly chapters, or part chapters, of the research project (at undergraduate and post-graduate level) are being used in assignments prior to the start, or even within the framework, of the research project itself. So, you may be required to produce a draft 'Introduction' chapter as an assignment before you actually start the research project. Also, you may be required to submit a draft literature review, or in some cases even the detailed plan for the methodology of the research project, as assignments, in advance of actually conducting the primary research.

Because of its size and the requirements of energy and time to complete, the research project carries far more weight than any other part of courses. Failing the research project at undergraduate level may lead to a downgrading of your qualification from a degree to a diploma and limit chances of studying. At postgraduate level, failure is not an option, if you hope to gain a Master's qualification.

Why do research?

As Long (2007) indicates, research is what the enquiring mind does. Most people are involved in some form of research almost every day. For example, finding your way to a new educational establishment by asking directions is a form of research. Asking how much food costs in a restaurant, is a form of research, because it involves asking questions. Asking your partner what they want to eat for an evening meal is research. Asking your mother, father, brother or sister what they want as a birthday present is research. However, these activities may not sound like the research that you will conduct in your research project. The similarities are that you are posing questions, with a specific purpose in mind and you are asking individuals (in research terms these are respondents) to answer them. Responses to your questions will give you information and this may allow you to make decisions and then act on the basis of what you have learned. In fact, much of your life is taken up with making decisions such as: what to wear before going to study, how to get to university, what to eat for lunch, what time to meet friends, when to do your studying.

Common sense?

Many of the decisions you make are based on having made them before, (in other words using your experience). You have made these decisions on many occasions, so you are likely to regard the reason you have decided as 'common sense' (see Brotherton, 2008). You may also believe that what you regard as common sense, is what others would also agree is common sense. But is this always the case? You may regard it as 'common sense' to walk to university when it is not raining, but catch the bus when it is raining; the main reason for both decisions being to be dry and avoid getting wet. However, you may have a male friend who always walks, whatever the weather, and a female friend who always uses the bus. Are they applying the same 'common sense' that you are using? Your male friend may indicate that the walk to university is part of a fitness regime, and the weather is not important, while your female friend tells you 'I travel with my mates on the bus.' So, it would appear they are applying their own 'common sense'.

'Common sense', as we call it, is therefore really little more than our own individual idiosyncratic behaviour (Brotherton, 2008) based on our experiences, and it varies from person to person. It is therefore not 'common sense' in terms of being 'shared' sense! Common sense is largely a set of beliefs. However, beliefs do not necessarily require evidence to support them. Using 'common sense' is therefore not necessarily a particularly useful way to understand what happens in the world and why it happens.

However, collecting evidence in a systematic way to help understand what is happening now and using this to make decisions about the future is what we refer to as research and should be a far better approach than using 'common sense'. Unfortunately, the notion that 'common sense' is more useful than research seems to be a view held by many who work in the tourism, hospitality and leisure fields. This is compounded by the fact that almost all of us have had leisure/tourism experiences so we can all claim, using our 'common sense' approach, that we are experts!

What is lacking in the 'common sense' view of behaviour in the world, is evidence, and in particular evidence that is accepted by all involved. This is because what we call an 'evidence-base', that all involved in can study and draw on to support or argue against, is likely to be a far more satisfactory way of making sense of the world than what has been referred to above as 'common sense'.

It is research that can and should provide this evidence base. Research is the process by which the evidence base is created. As a result of research, it should be possible for those involved to state, '*based on evidence*, this is what has happened'. This evidence may also allow us to provide *likely reasons* why certain things have happened, in other words *the causes* of what we have found out through evidence gathering and may also allow us to *predict* what could happen in the future.

The value of research

Unfortunately, many students view research as 'a necessary evil' (Brotherton, 2008) – it is something that they have to do, but really do not want to do. They have to do it because their tutor has told them to and anyway research is the key focus of the research project. So, often research is regarded as a chore, rather than a potentially creative and satisfying experience. This is particularly the case in the field of tourism, hospitality and leisure. The area is usually regarded as more vocational than theoretical, hence the link between conducting research to add to our theoretical knowledge is far less prevalent in the tourism, hospitality and leisure fields than, say, physics or chemistry (Brotherton, 2008). 'Learning by doing', 'learning on the job', rather than 'learning from theoretical research' has far more meaning in the tourism, hospitality and leisure fields. So, unfortunately, theory is often regarded as having little value or relevance to these related field. However, several other vocational subjects value research greatly. For example, engineering is usually viewed as an applied subject and strongly vocational. But it has a strong theoretical base and without this, developments such as the internal combustion engine, jet propulsion and electric car batteries, which have each been applied leading to major changes in global society, would not have occurred.

Many books and articles suggest research is all about generating new knowledge, gaining new insights and making discoveries (see, for example, Veal, 2011 for a longer discussion of this). This implies that research is always about discovering new things. However, your tutor or lecturers may have already indicated, or you may have read that your work is meant to build on and possibly fill 'a gap in our knowledge', or 'make a contribution to the existing body of knowledge'. This approach is a much more helpful way of thinking about your research project. You may discover something new in your research, but it is much more likely that you will work within fairly tight structures and not add much new knowledge. You are more likely to be repeating, or even replicating what someone else has done, but in a different setting or context, at a different time, with different respondents and possibly a variation of the original questions asked. After all, if you are making a contribution to knowledge or filling a gap, you must link your primary research very closely to what has already been done – as revealed in the literature on the topic. Although this means the type of research you will do is probably less exciting than you may at first have thought, the positive aspect of this is that it will be less of a challenge or hopefully more 'do-able'.

Your research is for a specific purpose – to gain a qualification – unlike much other research which is to 'push back the boundaries of understanding', (or to find answers to questions given to a paid consultant to answer). You should view your research project within its context, in that it has similarities with other types of research but has its own set of rules and a particular format as well as a key aim, which is to gain a qualification. You should be aware that part of the context for your research is that you have limited resources. Primarily, you are limited by the time you have to complete the dissertation, but also by the fact that you are not an expert researcher – you are learning how to be a good researcher, and this is where the qualification element is important.

When you conduct research, you will almost certainly be asking the following questions:

- What?
- Who?
- When?
- Where?
- Also, you may ask how many?
- But the most important question in terms of dissertation research is: Why?

Answering the question 'why' is attempting to find out not just what happened, who was involved and where and when this occurred, but the reasons behind the answers to the 'what', 'where', 'when' and 'how' questions. Answering 'why' is the key aspect of a student's research project and is part of the process of analysis. For this reason, the part of the project which attempts to explain the results, in other words provide reasons why specific results were obtained, is the most important part. Providing answers to the question 'why?' is also the reason that research is valuable. It helps us understand complex social activities and events. This understanding is not just useful in its own right, but can be employed in relation to future activities, in terms of further understanding, and is likely to help predict what may happen in the future.

Will research be useful for me?

Research will be useful for you beyond your student days because the process of collecting data and analysing it involves skills that are used in many jobs in the business field. Of course, some of you may go on to conduct further research after completing your project, so this research may be a stepping stone to future research. It is quite likely that many of you reading this will at some point in the future obtain a management position in the tourism and/or related studies field. Managers are continually involved in the process of collecting information via research to assist them in their decision making. A key role of any manager, but also those in non-managerial business positions is to prepare and then present reports. The preparation and production of reports will require the skills developed while involved in producing a research project of the type on which you are about to embark.

Scientific research

Although there are similarities between most types of research, there are important differences. The research you may be involved in is within the context of what is known as scientific research. This means it has certain characteristics and is conducted in particular ways.

According to Brotherton (2008) there are nine important characteristics of scientific research, and these are summarised in Figure 1.1.

Purposeful: research has a clear focus with specified aims and objectives.

Testable: the questions being used must be answerable. The main question (s) must be written so that it can be tested or proved. It is must be phrased in either a positive or negative way, so that it can be determined if the proposition can be supported or not.

Rigorous: sound, logical thinking is required to produce the overall research design and specific techniques.

Replicable: the research should be open and transparent, so other researchers should be able to repeat the research to test its rigour and the accuracy of its results.

Has precision: although the research is likely to be small scale and not as complex as most real-world situations, it should be the case that the design sample is a true reflection of the population.

One can have confidence in it: there should be a very high degree of confidence that the findings from the sample are correct.

Is objective: the research should be as free from bias and subjectivity as possible.

Can be generalised: the research should be applicable to not just one context, but generally applicable.

Parsimonious: the research should be focused, economical and manageable and capable of revealing the key factor(s) not just all the factors.

Figure 1.1: Characteristics of research (based on Brotherton, 2008)

■ Research in the social sciences

Brotherton's nine characteristics of scientific research are closely linked to the idea of natural science. Although Long (2007) suggests some characteristics of tourism research that can also be found in Brotherton's characteristics in the social sciences generally (and the field of tourism is within the social sciences), it may not be possible to conduct research which is entirely objective, is easily replicable and can be generalizable. In fact, a large number of social science researchers, including many in the tourism field, do not accept that their research can be objective (here objective means free from bias or adopting a neutral position). As Long (2007) argued it is probably not possible in the social sciences to have an entirely objective approach and he goes on to argue that maintaining the pretence of objectivity is likely to be confusing and unhelpful. As he stated: 'It is unreasonable and misleading to see the researcher as a neutral, sterilised research instrument' (Long 2007:3). However, Long attempted to indicate the key elements of what he termed 'good' research in the tourism field and stated it requires:

- A carefully formulated problem/issue/topic, with precise definitions and terms;
- An understanding of the context in which the problem/issue/topic is located;
- Credible data collection techniques from a suitable set of sources;
- Appropriate and competent analysis of the data;
- Informed interpretation, and a preparedness to acknowledge the possibility of contradictory evidence.

In summary, Long argued that good research is carefully formulated, the researcher knows where the research is located in literature and in its wider context, it uses credible data collection techniques and applies appropriate data analysis in a competent manner to provide informed interpretation of results, while at the same time acknowledging the limitation of these results.

A key aspect of a good piece of social science research (including your research project) is that it creates an argument that is based upon evidence. The evidence is used to state, something to the effect: 'on the basis of this evidence I argue these conclusions are true'. Therefore, when reading the piece of research to ensure that its conclusions can be accepted, it has to be the case that: 'the arguments in the report are valid, that the data do measure or characterise what the authors claim and that the interpretations do follow from them' (Sapsford and Jupp, 1996:1). Considering this statement, in relation to your research project, means that certain key aspects need to be clear, when it is read by your examiners. They will need to feel:

- That the claims and/or arguments you make are based in the evidence you have gathered;
- That the overall research approach and the specific techniques you used were appropriate;
- That your analysis of the data is actually based on the data that you have collected.

Self-awareness

Before embarking on your research project, you need to develop and apply honest self-awareness, which means that you need to be able to assess, as accurately as possible, your ability to create and complete your study. This self-appraisal is vital – many of you are not necessarily experienced at doing this and are very likely to over-rate your abilities! It should be based on evidence from written work you have done previously. It is probable, however, that your research project will be the longest, most sustained piece of writing you have attempted so far in your academic life! It may seem like writing several assignments at the same time. You should, therefore, not base your self-appraisal solely on a few written assignments that had a short-term framework for completion and were focused on a specific topic. Nor should you rely entirely on what you achieved under exam conditions in just a few hours of writing. Your research project will require effort *over a relatively long time*, so diligence, prolonged concentration and commitment are essential.

Part of this self-awareness also involves consideration of why you are doing a tourism or tourism related degree, in other words the motivational factors behind your study. You need to appraise yourself to indicate how keen and committed you are to your topic/theme or issue, as those who are really interested in a study topic are more likely to not only complete it, but also do well in terms of the mark received.

As part of this self-awareness, you should consider these questions and answer them honestly!

- How well do I cope under stress?
- How well do I respond to critical comments about my work?
- How will I respond when my supervisor criticises my work, directly to my face?
- How do I cope with/respond to deadlines?
- How will I respond if things do not go according to plan?

It is quite possible that what you have planned for your research project does not work out as you intended. For example, perhaps there are not sufficient visitors of the appropriate type to observe the behaviour of groups at the heritage attraction that is the focus of your primary research. Or using another example, perhaps attendees at the sporting event you want to study will not be interviewed because it is raining, or using one more example, not enough festival goers will complete your questionnaire as the band they want to see are just about to play! This means that you need a contingency plan – if you do not have such a plan, you may not have a research project at all. So, do not assume that people whom you plan to obtain information from in your primary research, will want to answer your questions. This does not mean that you have to bribe them to receive responses, but you at least have to show that you are keen to talk to participants and hope your enthusiasm is rewarded.

Think about the practicalities of when and where to conduct your primary research. A UK beach in winter is not a good time to conduct interviews with 'sun, sea and sand-seekers'! Ski tourism in much of Europe's Alpine area is not a particularly common activity in June! Hotels and restaurants in Europe are not usually as busy in October as they are in August! These may seem rather obvious 'mistakes' that it will be easy to avoid, but I remember a student determined to observe attenders, conduct interviews and questionnaire surveys at a UK five-day outdoor festival in June. However, although relatively unexpectedly, it rained on each and every day, with severe impacts on the quantity and quality of results!

You need to be aware that what interests you in your research does not necessarily grab the intention of your supervisor as much as it does you! You should also be aware that your examiners may not be experts in your chosen topic. However, what they are most likely to be is far more knowledgeable and experienced in *actually conducting research*. So, for both your supervisor and your examiner(s), your methodology may well be of greater interest than the actual topic you have researched.

■ Awareness about the nature of research

As part of the process of self-awareness/self-appraisal, you should ask yourself what you already know and understand about research as a concept and as a process. Again, reflecting on previous research experience, in whatever context, should be particularly useful. If you have done research (some form of research project) previously, this is likely to help with what you are working towards now, not necessarily because you have selected a similar topic, but because of what you learned about the process of conducting research. However, you should be aware that your current research project is likely to be much longer than other pieces of work and is in greater detail and depth. Although many student research projects are largely reproducing someone else's work in a different context, it also means you will need some originality and creativity.

■ Awareness of the requirements of the research project

You need to be aware of the institutional requirements of your research project, with regard to such matters as its length, the timeline including deadlines, and penalties for plagiarism. These details should be closely linked to the rules and regulations of the university in which you are studying. There will also be guidelines on what to expect from your supervisor and what he/she will expect from you, and this is discussed in more detail below.

You should be aware that each institution that awards degrees has different rules and regulations. Important dates, deadlines and the overall timing for your research project may therefore be different from other programmes at the same institution. If you are studying at a different institution from your undergraduate one, then do not assume the rules and regulations, important dates, timings and deadlines are the same as before. You will almost certainly receive documents about your specific type of research prject, with many institutions providing this in electronic form. Read it carefully, check it and ask questions where necessary. It is probable that your supervisor will go through documents with you, but there is no guarantee that this will happen, particularly with issues that have resulted from the impact of COVID-19, so you may need to ask your supervisor questions about the rules and regulations.

Awareness of the role of your supervisor

It is very likely that you will be allocated to a supervisor in your institution early on in the study process and usually by the time you have submitted a research proposal. Your supervisor will help and guide you as you develop the project, but please remember that they are not there to feed you information or give you references or tell you exactly what to write. You should ensure that you arrange regular meetings with your supervisor and agree in advance what work you will have completed and will discuss at the tutorial. However, you need to be aware that your supervisor is likely to be teaching during much of the time you are preparing and writing your study. Very research-active academics are also likely to be away at conferences and/or workshops when you are working on your study. Do not expect your supervisor to contact you frequently; it is almost always your responsibility to contact them and set up meetings/tutorials. Only when there are specific times that the department or faculty has set aside for special events in relation to the research project should you expect to be contacted by your supervisor. So, do not expect to see your supervisor without prior arrangement and if you cannot make a tutorial make sure your supervisor knows in advance.

Students are likely to believe that they are being judged on their ability to present results related to a particular topic, issue or questions and convince, at least initially, their supervisor and subsequently examiners that these have revealed something new. However, supervisors and examiners are at least as interested in how the student conducted the research. In other words, the different techniques used, and the overall research design and philosophy are of great interest and importance to the marker. This is not just because the way that the research has been conducted will have a major bearing on the reliability and validity of the results, but the marker wants to find that the student has learned something of the process of conducting research at this level. The supervisor should advise on, and the marker will want to see, why a certain approach was adopted, why particular techniques were used and when, as is possible, things did not go to plan, what changes were made how these were linked to a contingency plan and the impact this had on the results of the study.

Case Study 1.1 provides material on the role of the supervisor and is based on documents produced by a UK university, intended to be given to students just embarking on their research project programme.

Case Study 1.1: The role of the supervisor

These guidelines are intended to clarify what your supervisor expects of you and what you can expect of them. They are also intended to promote equity, so that all students can expect a similar level of support. Please bear in mind that they provide guidance and do not set out rigid rules. As in any working relationship there is some scope for negotiation. Your supervisor wants to support you in your work and in return expects you to treat them in a considerate and courteous manner. Failure to attend booked tutorials, failure to produce work as promised or late submission of work for review will mean that a request for extra help is much less likely to be agreed to. Remember the help you get depends on what you are willing to put in. The responsibility to produce a good dissertation is yours, not the supervisor's. Finally bear in mind we are all busy and we may not be available during the vacation time to read your work or hold tutorials.

What you can expect of your supervisor!

Your supervisor will:

- Meet regularly with you at prearranged times
- Support you generally in your work on the dissertation
- Offer advice on the topic generally, work programming, making use of the literature, research methods and analysis
- Review and advise on draft questionnaires and other research instruments
- Review and advise on draft chapters

Your supervisor will not:

- Give a tutorial unless the meeting is prearranged

- Normally give a tutorial unless you have previously submitted, in good time, written work to discuss
- Provide a detailed list of references for your literature review
- Review work that is submitted only at the last minute
- Rewrite questionnaires or other research instruments
- Make detailed corrections to English and grammar (although they will tell you when there are problems)

What is expected of you!

You should:

- Show a consistent high level of commitment
- Create a work programme and work to it
- Provided written work for review on time
- Agree regular meetings with your supervisor and attend them
- Plan ahead for the tutorial meetings with your supervisor rather than try to arrange them at the last minute
- Make full use of library and internet sources as well as relevant librarians in your search for secondary sources

You must not:

- Expect to be able to arrange an instant tutorial
- Normally expect your supervisor to provide research contacts for you
- Send out surveys or approach any external organisation without first consulting and obtaining the agreement of your supervisor.
- Expect a tutorial unless it has been arranged in advance, as your supervisor will be very busy teaching on other courses. It is always your responsibility to maintain contact with your supervisor and to ensure you have regular meetings.

(Based on documents from the University of Westminster, UK)

Awareness of how you write

A very important aspect of self-awareness is knowing how you write your academic work, and it is best if you are aware of this before you actually start writing your study. However, you may not ever have thought about this before, particularly if you believe that we all write in exactly the same way. Nevertheless, this is not the case. A brief anecdote from my own experience should be sufficient to indicate what I mean here. I worked at a university in New Zealand in the 1990s. On one occasion, with several other early/mid-career lecturers, I attended a one-day workshop, run by senior academics, on how to write to improve our chances of getting articles published in top quality journals. One academic gave her presentation on writing styles and asked for comments and questions at the end. After her presentation, she was greeted by silence and bemused expressions from the audience. In her presentation she had indicated that she planned the article, then read literature, thought for a while and then produced elegant, well-formed sentences, almost none of which had to be modified. She then said that she reflected on what she had written, read literature again and then carried on writing and, once again, created almost perfect sentences that only rarely needed modifying. She also stated that she started her writing at the actual beginning of her article and carried on writing, in the chronological sequence that she had planned, until she reached the end.

The presenter was greeted by silence, because, as her audience gradually indicated to her, no one else at the workshop wrote in this way! The great majority informed her that they needed to write many drafts, often up to twelve before feeling they had achieved a satisfactory quality. Also, some of the audience said that after writing several drafts, they frequently changed the order of paragraphs and sentences from the early drafts. Yet others said that they often had to first write several drafts, which then had to be all thrown away, before, as they indicated, they started to write something they felt confident about. You may be (lucky) like the presenter at the workshop, or probably, more like almost everyone else at the workshop.

Although all of you who are working to produce a research project can write (you have probably achieved academic recognition through you writing already), your ability to write has been learned over time. One way to think of writing is a craft – (although some people may consider it perhaps as an art). If it is viewed as a craft, like other crafts, writing is usually learned in stages.

Writing should be regarded not as a single act but a process. It should also be remembered that writing is actually writing, not preparing to write. It is only when the words appear on the page that the mental activity of authorship begins at all. You may have heard other students, particularly Masters or PhD students, say words to this effect: 'I have finished all my research, now I just have to write it up.' This implies they have done the hard work and the easy job of writing (the dissertation or thesis) is to follow. Nothing could be further from the truth. 'Writing up' is what doing a dissertation or thesis really is!

Unlike the presenter at the workshop discussed above, for the great majority of authors, much writing is in fact re-writing. As Watson (1987) indicates, most good thoughts are afterthoughts; they come when a first draft has 'woken the mind up' and it therefore needs no further prompting. Watson goes on to claim that it is in the mental state between first and second drafts that 'real intelligence' begins to appear.

The following are ways to link quality of writing with the audience for what is being written:

- A student essay can be written for a captive audience, usually an audience of one.
- Writing a thesis is writing for the world, the audience is more than one and the thesis should be of publishable quality.
- Writing a dissertation is somewhere between writing an essay and writing a PhD thesis. The audience is more than one – it involves supervisors and examiners. The quality should be of high enough level to have parts, particularly if writing a Masters dissertation, adapted for publication.

It is worth remembering that most of us cannot speak as well as we write. This is because we have time to prepare what we write, while speaking is often spontaneous and we do not have as much time to think when we speak, as when we write. Those who do speak well have often memorised something that was originally written! According to Watson (1987) writing is like a cow 'chewing the cud'– this is the cow regurgitating what is in its stomach and chewing this again to break it down! What Watson means by this is that it is a gently ruminative process – in other words an effort of slow diligence and steady perseverance.

The nature of some writing is however very different from other types. A birthday card to your mother, an email to a friend, or a text message, or a message on Facebook or WhatsApp, will require a different writing process and be in a different style to an academic essay. Yet it is easy to forget this when you sit down in front of your computer to write your dissertation. You need a formal style when you write the dissertation, and this requires effort, concentration and different skills to those that you use when you write that informal message to a family member or friend. This formal language will nearly always take longer to 'get right'.

A dilemma for some students, when writing their study, is the need to be original on the one hand, and the requirement to relate what you are writing to existing literature. Therefore, it is important to remember that your supervisor and, eventually, your examiners will not be interested in your views, merely because they are yours. You may have what you regard as an original contribution to make in your study, but you must go through the formalities of locating your study in relation to the appropriate literature, before you can claim that you have found and filled the gap in the literature with your work!

So, you need to present your ideas lucidly and in 'correct formal dress.' You need to be very aware of the climate of opinion in which your thoughts exist, and have a clear professional regard for the existing state of play – in other words you must know the key themes, ideas, issues and concepts related to your project topic and the major literature where this material is located. However, I have referred to writing as a craft above, so you should also remember that you are an apprentice researcher. Apprentices often make mistakes, and the apprentice may be forgiven for making mistakes. Mistakes are also correctable – this is part of the learning process. What supervisors and examiners are unlikely to forgive is if you do not seem interested in trying to correct your mistakes and get it right!

If you can provide an account of how your research developed over time, this will be of particular importance to whoever reads your study – it will be telling a story. In this account, you should give the rationale for why you did what you did and what changes were made, why and with what effects. Once again, when telling this story, it is important to remember that you are an apprentice researcher. If you were an apprentice chair maker, you could show your teacher the finished chair, the product of your efforts, and explain how you cut the wood and then joined it together – you could also sit on it and show that it works as a chair! It is not identical when conducting research, but you are an apprentice – which means there are some important similarities:

- You will have designed the chair first;
- You will have selected the wood and the tools to make the chair;
- You may have made a model of a chair before the real thing;
- You may have tested different materials and possible tools;
- You will learn by doing (i.e. actually making the chair);
- You will have learned something by actually making mistakes;
- Your main physical result is 'a chair';
- Any more chairs that you make after the first will probably be better than the first.

Very importantly, in terms of this process, you should be able to explain what you did, why you did it the way you did, what you have learned and how you could improve the process in the future! The nature of this explaining is very close to the processes involved in your study and particularly similar to the analysis part of it.

Case Study 1.2 is concerned with formal academic writing and indicates the way that I write. It may not be the precise way that you write, but you should note that what I write is at least at the same level as your study. It may also be important for you to be aware that I write while sitting at a computer, which is probably the way that you also write, but thirty years ago when I did my Master's thesis, I hand wrote it, before it was typed up!

Case Study1.2: How I write

Sitting in front of the computer (the numbers indicate the process in chronological order):

1. Brainstorm: put down ideas as they occur, not organised, fairly quickly, so as not to lose/forget ideas
2. Read and reflect on ideas and attempt to organise
3. Re-organise ideas
4. Write a first draft
5. Read, reflect and edit this draft
6. Write a second draft

   (and so on through several drafts, if necessary)

But it is not this simple as

- Brainstorming continues throughout with new ideas added.
- Time is necessary to write, but also to reflect before continuing the organising and writing process. The 'let's sleep on it' mentality is important.
- The time gap between stages will affect where writing is 'picked up' again and probably the final product.

Case Study 1.2 indicates a circular, or more accurately, a spiral process. However, as the case study suggests, it is not quite as simple as this. One important aspect of the case study for you to be very aware of is that time is critical. Because I (and many other writers) do not get it right first time, it takes a good deal of time to produce work that I am happy with!

Writing this chapter in your study

The Introduction chapter in your research project serves a number of different functions, as indicated below. This chapter:

- Indicates why the topic you have selected is important in relation to your specific field of study – this the rationale for the study;
- Provides an overview of the important literature within which your topic is located;
- Outlines the aims and objectives/major research questions of the project (these are discussed in detail in Chapter 2);
- Provides an overview of the structure of the project, with a brief summary of what is to be found in each chapter.

It is a good idea if the structure of this introductory chapter is in the sequence of the bullet points above, with the rationale provided first, in other words an explanation of why is this a important topic, followed by brief discussion of the key literature (that you will discuss in detail in the Literature Review chapter) to support the rationale. This can then be followed by the specific aims/objectives/research questions. Finally, the chapter can indicate the structure of the rest of the project, briefly outlining the content of each successive chapter. The length of the introductory chapter is not set in stone but should probably be around 10% of the total length of the study in words/pages.

Students may believe it is vital that they write the introductory chapter first – after all it will be the first chapter that the supervisor and examiner read. However, although it may be a good idea to draft out a first chapter early in the project write-up stage – and for some Research Methods courses in universities an introductory chapter may be an assignment – the final version of the introductory chapter is probably best left until almost the last thing that you do. The major reason for this is that if you leave the Introduction to last, then you know what is to follow in the main chapters that follow it, as at this point you have written them! In other words, you will not write something to the effect: 'Chapter X contains ... and Chapter Y is concerned with....', when in fact this is not the case as you have modified these chapters since writing the original Introduction!

It is not a good idea to tell your reader (the examiner) that certain things are to be found in particular sections/chapters in your study when they are not there. Although most examiners like to give students the benefit of the doubt and are looking to give marks where they can, it is only human nature to get confused (and probably annoyed) when something is not as stated. It is also likely that the examiner will believe that the project was rushed or left until late in the day, if the content, as stated in the Introduction, is not what is actually found in the main chapters of your study.

If the Introduction does not provide an overview of what is found in later chapters, it is also likely that the examiner will believe that the study was rushed. Examiners are also used to reading material, in which it is relatively easy to work out that the student does not really know precisely what they have written, or really understand what they have written – but are nevertheless aware that they must put something in a specific section! A key message here is what I have indicated above: *writing is largely re-writing*, and you may need to redraft this chapter several times. This clearly requires time, and this is why you have been given months and not merely a few weeks to write the project.

Student activities

1 Produce a brief summary of the research that you have previously conducted and indicate in one column the 'highlights' of this and in another column alongside it, any problems you encountered. In relation to the problems, indicate how you overcame them; and in relation to the highlights, indicate why you believe they are highlights.

2 This chapter has included the following questions:
   - How well do I cope under stress?
   - How well do I respond to critical comments about my work?
   - How will I respond when my supervisor criticises my work, directly to my face?
   - How do I cope with/respond to deadlines?
   - How will I respond if things do not go according to plan?

   Answer these questions - honestly!

3 In relation to Case Study 1.1, indicate what you expected in connection with the supervisor's role and what you did not expect in terms of the role of the supervisor? Having read this chapter, how do you now view the role of the supervisor?

4 Put yourself in the role of a supervisor who has not received promised work from a student who then turns up late. What will you say to the student?

5 In relation to Case Study 1.2, compare what I have indicated about the writing process with the way that you write. Construct a table and put in your comments use the headings: 'Similarities' and 'Differences'.

6 What are the possible implications for the way that you write, for the writing of your research project?

2 The Literature Review

Introduction

Although your study has an introductory chapter, the first really major chapter in it is very likely to be the 'Literature Review'. This chapter will require a good deal of preparation in terms of reading and probably, depending on your writing style, a number of drafts, before it is good enough to be included in your final document. The nature of the literature you should use in your study, is discussed in the chapter that follows, and it indicates how the review should be written and also how to structure it.

What is literature?

This may seem an easy question to answer. Clearly, there is a very large amount of literature that has been produced over time and continues to be produced. Your initial reaction to the question above may be to say something like 'literature is found in books and other printed works and is made up of words'. But it also includes tables, figures, charts, graphs, diagrams, photographs and maps. So, you could summarise this as 'anything that is written'. However, there is more than this, because literature also includes radio, TV, film, DVD, and of course the internet, so there is a visual and audio component, which, for some types of researcher, is very important. So, it is probably better to think of literature as information. Also, literature in relation to your study is information in the public domain. You can therefore find the information in books and other printed material, in libraries, on the radio, TV and film, but also, increasingly, on-line.

■ On-line resources

Traditionally, students doing their research project would spend a good deal of time in the university library or learning resource centre. This is no longer the case in many countries. Students are more likely to be on-line and the time spent on-line has increased since COVID-19 when physical access to libraries has been limited. Fortunately, most of the good academic sources for relevant literature, in particular journals, are available on-line and can be accessed remotely. Your university/college will advise you on how to access journals and other library-based material, both within the library and remotely.

For many students, the first place they will go to obtain information is the internet. However, you should be very aware of (and wary about) problems and limitations of internet-sourced material. In terms of preparing the research project, you are meant to be using predominantly academic sources. The main reason is that academic sources are those that can be trusted to be more reliable than other sources. Before being published in a journal, an article will have been read and reviewed (or critiqued) by at least two, or frequently three academics, who are experts in the subject matter of the journal article. These reviewers will suggest any changes they see as necessary, such as adding material or removing material. They may indicate that there are inaccuracies in the material, or something is missing. The review process is meant to be 'tough' and many submitted articles do not get published; those that do usually have to be revised and sometimes substantially revised involving several submission and re-submissions. Hence, journal material is the most *authoritative* that you will be able to access.

Academic books and book chapters may also have been through a similar review process to journal articles, but this does not always occur. Other books may have been through a critical reviewing process, but they may not have been, and it is not always possible to tell. International bodies such as the UN and the EU, national governments, government bodies such as ministries, non-government organisations, charities and private organisations produce reports and similar types of publications. These can be authoritative, if based largely on reviewed academic sources. However, if produced by governments, they may be little more than propaganda. Newspapers and magazines do not go through the same review process as academic journals articles, although there is usually some editorial control. TV and films do not have the same review process as books, although, as with newspapers, there is editorial control.

How authoritative is the material?

Remember, anybody can put anything on the internet! Wikipedia is often thought of as an authoritative source of material. However, it may not be authoritative. The quality depends on who is writing the material. Some of the Wikipedia entries are excellent and of a high standard, others are not.

Remember that famous celebrities have often written their own material for Wikipedia and change it when they feel like it! Do not assume that what you read on the internet is accurate or true.

Another very important factor is that it is (too) easy to plagiarise from the internet. Simply 'cutting and pasting' can be too tempting for some students. Remember, most dissertations have to be submitted through computer software such as 'Turn-it-in'. Those caught cheating (plagiarism is cheating) in this way face very severe penalties, usually meaning they have to re-submit their research project, or do another one. In extreme cases of plagiarism, students can be thrown off the course and out of the university.

Below is a summary of major sources of literature for your dissertation with the most authoritative at the top, the least authoritative at the bottom.

- Academic journal articles (most of which can now be found online)
- Academic books and edited book chapters
- Non-peer reviewed academic material/other non-academic books
- Popular media (newspapers, magazines, TV, radio, film)
- The internet (non-journal related).

The rationale for the literature review in your study

Some students believe (wrongly!) that if they are in the process of discovering something new, then there will be little or no literature that it is relevant to their study topic.

Remember that your study will (hopefully) make two types of contribution: one of these is the practical application of your findings, the other contribution will be the intellectual one. This second one is the contribution to literature. It is this second point that is particularly important in relation to the rationale for a literature review in your study. You need to indicate what the literature is at the current time, to show where you have made your contribution.

However, in relation to the fact that the study is to be assessed, you also need to demonstrate that you are very aware of the key writing and important comments about your topic. Your supervisor and your examiners need to know that you are aware of the major literature, as through this, you demonstrate that you are knowledgeable about the topic you have chosen. Without a detailed awareness and a clear understanding of the important literature, it is very difficult to make a claim about your 'contribution to knowledge', or 'filling the gap' in the literature. In addition, a good literature review gives authority to your study (Long, 2007). It means that you should have established your ability and demonstrated your expertise, which should impress your supervisor and, in particular, the marker of your study.

Your research in the context of the literature

A key purpose of the literature review is to indicate that the topic, issue or problem that you have decided to investigate really is worth researching (Thomas and Nelson, 1990). The way in which you do this will be to demonstrate your knowledge and understanding of the past literature in relation to your topic, issue or problem.

In relation to the topic on which you have decided to focus your research, you may have read that certain aspects are controversial. In other words, there may be a lack of agreement on a specific aspect of your issue or topic. This could form the basis of your research project. If you select this controversial aspect of your topic, then it is relatively easy to argue how you are making your 'contribution to knowledge' – it should be adding to the arguments about the topic and will hopefully be providing evidence on one side or the other of the argument. Selecting a controversial topic also provides you with a good rationale for studying it in your project, as even in a small way your research could add to discussions about the topic

Whether or not you have selected a controversial aspect of an issue, your literature review should be written in such a way that it builds the case for the research into the specific issue/topic/theme in which you will conduct primary research.

The wider field

You also need to show in the literature review that you are aware of literature that you do not discuss in any detail. This may sound a rather pointless activity, but, in fact, is important for you to show that the topic is located within a broad field of your selected tourism, hospitality or leisure field and this is part of the rationale for studying it, but also that you have not discussed this particular literature, because it is not strictly relevant to your dissertation. This is very much part of ensuring that you do not review everything that may appear relevant (see Case Study 2.1). But mentioning and then not reviewing the particular piece of literature (and saying why you have done this) can be a very important part of writing the literature review.

It is part of a focusing down process, but also reveals that you have read widely and can discriminate between the relevant and the not so relevant, which will be an important part of what your supervisor and examiner will be looking for when they assess your work.

If you really believe that there is little or no literature on your chosen topic, you need to be particularly careful. As indicated previously, students often claim there is little literature on their topic, but this is usually because they have not spent enough time searching in the library or on-line. Artificial Intelligence (AI) is an example of a relatively new topic, and it is being applied in tourism and hospitality. If you decide to analyse its use in hotels, and select one hotel as a case study, you may find that your selected hotel is just beginning to use it, so there is little literature on the specific issue of AI and your hotel. However, this does not mean that there is not a significant amount of literature on: a) AI itself, b) the use of AI in hospitality and tourism, and c) the use of AI in hotels. Hence it is not accurate in this example to tell your supervisor 'There is no literature on my chosen topic'. You will not get a good mark if you claim that there is little literature, when in fact there is. Your examiner is likely to think: 'This is a lazy student' or 'This student did not really understand what s/he was researching about, so it is not surprising that s/he could not find relevant literature'. If there really is little literature, you may believe that you are working in a new, exciting under researched area. However, your supervisor and examiner may have a very different interpretation. If there is little or no literature, it could well be because your topic is trivial and not worth researching about at the level of your study. It may be, in fact, so low level and obvious, that it does not merit academic study!

Although your literature review, particularly if it is a deductive research project (see Chapter 3 for a discussion of deduction), will probably appear early in your work, it is not to be regarded as a 'standalone' part of your study. A good literature review will be used later in the project. After reporting your results in the analysis section of your primary research in the project you should be able to refer back to some of the literature you have discussed in the review chapter. This is not always understood by students, who believe that once they have written the literature review, there is no need to refer to it again. However, what makes a good, if not very good, research project is when a student refers back to several pieces of literature found in the review. This may be to indicate that the findings from the primary research are similar to the work referred to in the literature, or possibly the results are different from this literature. It is this process of comparing your results with the literature that those who assess your work are looking for, as this referring back is evidence that you know the literature on your topic well. Through a process of comparison with literature, this also indicates how your own primary research relates to the literature. Making comparisons is a key element of analysis and this is a major activity in the chapter of your study which focuses on discussion of your results.

If your study is in the positivist paradigm using a quantitative approach (see Chapter 3 for a detailed discussion of these terms), the literature review is likely to be very significant in demonstrating the theories that already exist in relation to your topic. It is also important that you indicate where any hypotheses (see Chapter 3 for a detailed discussion of this concept), that you have created have come from in terms of the literature.

■ The methodology and approach

When discussing studies from literature that are very relevant to your chosen topic, it is a good idea to give some details of the research methodology and techniques used to conduct the research. Hence, you may discuss a piece of research that used a specific technique (such as a questionnaire survey) and this was carried out using, for example, a face-to-face approach, in a particular geographical location and was conducted using one type of sample (e.g. a purposive sample). One important reason for giving this detail is if you intend to use a similar approach in your primary research. One of the key defining factors, particularly for deductive, positivist research (see Chapter 3 for a detailed discussion of these terms) is that it is replicable. Hence, you are allowed to attempt to replicate someone else's research approach, and this will not be considered as plagiarism, providing your research is not identical. However, it should be relatively easy to make your research different by, for example, having a different location to conduct the study, using a different type of sample, doing the study at a different time of year, or by modifying some of the questions from the original study. If you modify some of the questions from the original study, but retain others, then you will be in a position to directly compare your results to the questions in the original article upon which you are basing your primary research.

If you want to replicate much of this original study but intend to have an important new approach, your rationale for this will need to be clearly explained. It could be that you are arguing that the reason the original study achieved the results that it did was because of, for example, the research technique, or the sample, or the location and timing of the research. You will be in a position to argue that either your research will come up with very similar results to the original study, or possibly different results. In fact, you may be in a position to set up hypotheses, based on the original study, to be tested and confirmed, or rejected, by your study.

Yet another reason for indicating how the research was conducted is that you are not intending to replicate a particular study but wish to have a completely different approach. In your reading, you may have found that the particular topic you have selected, has nearly always been researched most often via a closed-ended questionnaire survey. However, you have decided that you intend to use a qualitative rather than quantitative approach (see Chapter 3 for detailed discussion of these two concepts) via focus groups. The rationale for your approach, you may argue, is that the topic requires greater exploration than is possible using a questionnaire survey and you believe discussion in your focus groups will lead to in-depth responses and to greater insights into your chosen topic.

If your dissertation is in the phenomenological paradigm, being qualitative and inductive, (see Chapter 3 for a detailed discussion of these terms), you may have initially designed the research and then shortly afterwards collected data. Therefore, you may believe that the literature is not really important. This is not the case! In your research process, having collected data, then you should have compared this with what is already known – and this will be found in literature on the topic. It is likely, even in qualitative, inductive research, that almost all researchers will have a reasonably clear idea of what they could find out in their current research project – this is because they have conducted very similar research before and, therefore, they know most of the important literature on the topic. Hence, the literature review is just as important in a qualitative study.

However, if you have used an inductive approach, the location of your literature review will not necessarily be at the early part of the dissertation, as in a deductive positivist dissertation. If you collected data first then your literature review could be presented after the 'Results', but most likely before the analysis.

How do I know if the literature is relevant or not?

This is a common question amongst students. An important reason for the question being posed is that many students will feel that, until they have done a good deal of reading and thinking about their topic, they do not really know that much about it. At the beginning of the dissertation, they may feel that they know very little, so deciding on what is relevant and what is not, will not be easy at this earlier stage. The easy answer to the question about relevance, however, is 'You won't know until you have read it!' Nevertheless, there are some ways to find whether literature is likely to be relevant and therefore without having to read vast amounts of material.

Using abstracts

As indicated above, articles in journals are likely to be the most authoritative sources for your dissertation. Hence, the following section discusses how you can make decisions on whether a journal article is relevant or not. Articles in journals almost always have short abstracts. Abstracts are usually provided free online although you may have to pay for the whole article. However, most universities should ensure that you have free access to online journals as the library will have an institutional subscription. Nevertheless, an abstract is a summary of key aspects of the whole article, including the focus of the article, how it was researched and the major findings. Although they are brief, abstracts can be very useful, as they usually give details about research methodology, the location and timing of the research, the sample size (if relevant) as well as the major findings.

Key tests of relevance

To assist with revealing whether literature is relevant or not, you need to evaluate it. To do this you can use a number of key terms/headings or, as indicated below, questions in relation to the article that you are reading:

- What is the underlying research philosophy? Is it positivist or phenomenologist?
- What is the research design? Is it deductive or inductive?
- What type of data collection technique(s) were used? For example, a questionnaire survey, or interviews or a focus group?
- More specifically, if a questionnaire, was it closed-ended, was it conducted on-line, face-to-face or another approach? If an interview, was it structured, semi-structured or completely open-ended?
- What is the nature of the sample (if used)? How was the sample created?
- How has the data been analysed? Was this done manually or using a computer package? Was statistical analysis conducted?
- What are the major findings?
- Are the conclusions based on the primary research? Are they reliable and valid?

Although this approach may appear a time-consuming task when applied to everything you read, it is likely to be very useful. In particular it should help you decide which articles are more relevant and which less relevant for your research. Once you have used the questions presented above a few times, it should be almost a natural process to apply these to a new source of material and hence by then you will be much quicker. Usually, you will have been given a good deal of time to write your dissertation, but much of this time has been provided so that you can read and reflect before writing.

Using the questions should also allow you to compare and contrast articles (and other literature where relevant), see similarities and patterns emerging, as well as reveal gaps in the literature which you may be in a position to indicate you are trying to fill with your own primary research. When you are involved in the actual writing of the review, using the questions above in your evaluation of articles will also make your work less descriptive and more critical. And remember, the literature review should be a critical review.

Case Study 2.1 discusses an important theory in tourism and related fields. It indicates how this theory could be used in different research circumstances, such as an application of it to a specific destination, and a modern critique of the theory.

Case Study 2.1: What to include and what to leave out in terms of the literature review.

One of the major theories in the field of tourism is Butler's Tourism Area Life Cycle Model (often abbreviated to TALC), sometimes referred to as the 'Destination Life Cycle' model. This was first published in 1980 and a large number of tourism writers have tried to apply the model to particular locations and destinations where tourism is important. It is also a topic that some Masters' students have attempted to write about. In his article published in 1980, Butler acknowledged he was directly influenced by the works of important researchers, and, in particular, Christaller, Plog, Cohen and Doxey. As Butler indicated, he was also influenced by marketing theory and specifically the 'product life cycle'. His model was an innovative combination of existing tourism theory and marketing theory.

If your dissertation is focusing on how tourism destinations develop and change over time and/or how to plan and manage tourism destinations, you would certainly need to refer to Butler's (1980) model. But would you need to refer to those who influenced Butler (as indicated above)? The answer is probably 'No'. However, if you wanted to claim that Butler had misunderstood the importance of one of the theorists he drew on, Plog for example, then you would have to discuss Plog's theory as well.

With the type of dissertation research indicated above, it is much more likely that Butler would be your foundation theory and you would then examine more recent theories and apply these to your specific dissertation focus. Many of these more recent theories have provided a critique of Butler's theory, usually via case studies involving primary research, so this would give you important material for your literature review with which to compare your chosen location.

How to take notes

There are several reasons for taking notes while you read literature. Be aware that notes are not meant to be merely copying the original – this is plagiarism. When you read the literature, note taking should be your attempt to summarise what you have read. One important reason for doing this is that you will be putting what you have read into you own words. By doing this, you will demonstrate to yourself that you understand what you have read. Also putting it your own words greatly reduces the possibility of plagiarising. However, what may appear a less obvious aspect of taking notes, can in fact be seen as a major reason for doing so. To be able to take notes you have to be concentrating! If you are concentrating, it usually means that you will understand what you are reading, but if you do not understand, then the fact that you are concentrating and taking notes should lead to you re-reading and attempting to find out what it is that you do not understand! Note taking does not necessarily mean writing on paper but can be done on your computer, if you prefer to do it that way. This approach has the advantage of removing the need to type up your handwritten notes and it may also be possible to modify your typed notes into words and sentences that can be used in your dissertation.

■ Recording the sources

Also, when you are talking notes a very important thing you need to record, is the source of material. It is all too easy to get carried away with your note taking and fail to record the exact source of literature. You should be aware that supervisors and examiners become very annoyed if your references are inaccurate. In my experience, far too many students still fail to reference properly. Common mistakes include dates missing or incorrect, references not in alphabetical order, an inaccurate order in the wording in the reference and authors' initials, or first names used when they are not required.

To avoid having to take too many sets of notes, you need to be able to make decisions fairly quickly about the material you are reading. You want to know how the particular piece of literature relates to what you have already read and what it can contribute to your research. Remember that you will need to decide whether the material is relevant at all (see the previous section) and if it is not, then you will not need to read it and take notes, as this will be wasting your limited time. One way to avoid taking too many sets of notes is to photocopy articles or save the full text of electronic material. However, saving/copying material is not the same as actually taking notes. Photocopying is not a substitute for reading the material and putting a copy away in a file will not reveal much that is useful for your dissertation!

■ Organising your notes

Almost inevitably you will have to leave out a good deal of what you have read and made notes on. This is annoying, but remember it is always better to have to edit out material, than add in new material. According to Long (2007:28), these are the key considerations for you to use when you read and record:

- Put the research studies into some form of classification that groups similar studies together.
- Put the material in a logical sequence.
- Look for links and relationships in the material.
- Explain and evaluate the factual matters, the links and relationships and consider how they may contribute to what your primary research is intended to be.
- Compare what different researchers have found out.
- Identify not just the gap in the literature - 'the what'- but how the next stage of research should be conducted.

Selecting your topic

This is often a difficult task. Some students will find that a part of their taught course has inspired them and they are able to come up with a topic relatively easily. Other students may have a long-standing interest in a particular theme or issue. However, there will also be very many students who initially have little idea of what they want to conduct research into.

It may be a good idea before deciding on a specific topic to think: 'What is it that really interests me in the field?' A key issue here is that you will need to maintain your interest over a period of several months – the time that it takes to complete your dissertation. If you do not maintain your interest, there is the problem that you may not complete it, or your declining interest and commitment leads to a reduction in the quality of your work.

There are a number of 'favourite topics' that certain individuals who are actively involved in the tourism industry have researched. Brotherton (2008) provides a list of twenty-one major topics and fifteen lesser topics that one major group, tourism and hospitality managers, have researched. The most popular in this list are 'market research', 'new product/new service development', 'customer survey', 'service quality', 'benchmarking', 'the effectiveness of advertising' and 'site location feasibility'. This list suggests that marketing/advertising and quality issues are very important topics for managers. It is quite likely that these 'favourite topics' have literature to support them and, as indicated above, it is very important that there is a literature base for your research.

To help in making your decision, it is well worth considering what other students in the past have researched. This means you should look in the library, or in electronic databases, for students' work, in your institution, for at least the last five years. You could also talk to friends and other students to find out in what areas, in the field of tourism, they are interested. Table 2.1 indicates the provisional dissertation topics in three British University Tourism and Hospitality Department, as proposed by students at the beginning of the dissertation process.

Table 2.1: Examples of Provisional Dissertation Topics in three selected Tourism and Hospitality departments (2014-18)

- Evaluating the relationship between staff performance and recruitment processes in five-star hotels
- Service quality and the role of staff training programmes
- Social media in tourism: the value of user-generated content
- Success factors in ecotourism: what criteria should be used?
- Ecotourism and sustainability in Nigeria
- Possibilities for sustainable tourism in Malaysia
- Different types of hotels and the variety of services offered
- The motivation of visitors to dark tourism sites
- The effectiveness of integrated marketing communication strategies upon attendance at a UK international carnival

- The effect of management (TQM) on employee and customer satisfaction in the hospitality industry
- Event planning: a case study of wedding decision making factors in the UK
- The impact of artificial intelligence and robotics on the future workforce of hotels
- 'Cloud' computing technology in hospitality and tourism
- 'The full moon party' – what are the social impacts?
- Achieving effective marketing in food tourism in Ireland
- Food trails: their potential for tourism
- The motivation of aircraft cabin crew
- The impact of tourist behaviour on mass tourism destinations
- The impact of terrorism on global tourism: selected case studies
- Visitor attitudes to heritage attractions in the UK and Europe
- Collaboration and partnerships in the promotion of heritage tourism
- Tourism as a tool for poverty alleviation in developing countries
- Developing and evaluating a brand image model
- The role of national tourism organisations in promoting tourism
- The role of airport hubs: a comparative study in Europe and Asia
- How relationship marketing affects the hospitality business
- Tourism in small island development strategies
- The Thai diaspora and tourism
- The impact of tourism on turtle breeding on a Greek island beach
- A critical appraisal of film tourism and sustainability
- Motivations for millennials' participation in voluntourism

As Table 2.1 indicates, there is a very wide range of topics that can be selected, as well as geographical locations to choose from. The titles do not reveal the entire scope of the studies either. However, one very important factor, which is discussed in the next section of this chapter, is the scale of the study. In summary, the scale of study has to be manageable, given the time and resource constraints you have, so in general your study's primary research element, at least, should be small scale.

The scale of the research

On (too) many occasions I have heard students tell me that they cannot find enough literature on the chosen topic. Only very occasionally has this actually been the case. Usually, it means that the students have not searched hard enough for the literature which is the wider context for their research. For example, a student I supervised wanted to investigate the attitudes of visitors to a specific theme park in the UK. She told me that there was no 'academic study of this theme park'. I told her to look for the following:

- Generic literature on theme parks – there is a good deal within the context of visitor attractions.
- Research on attitudes of visitors, particular in relation to visitor attractions, including material on how to conduct such research,
- Literature on customer satisfaction and how to research this area,
- If the theme park did its own surveys and, if so, whether there were questions to do with visitor attitudes and satisfaction levels.

Often in trying to formulate your topic your initial thinking may be very wide. For example, you might think 'I want to do something on marketing of events'. After considerable thought and reflection, this can then be refined and focused until it becomes a manageable, 'doable' project with a title such as 'An evaluation of the effectiveness of internet marketing for the 2021 X Music Festival'. Another student might be thinking: 'I want to do something on quality in restaurants/bars/public houses'. After the process of refining and focusing this could become 'An investigation of service quality at three gastro-pubs in City X'. The 'refining' process is explained in more details below.

These two examples in the paragraph above, and that of the student who wanted to research visitor attitudes at a theme park also reveal two key aspects of research of this kind and these are features of most dissertation research. Each has a *conceptual* aspect and a *contextual* aspect (see Brotherton 2008). For example, the conceptual aspect of the idea of my student who claimed she could not find relevant literature was 'visitor attitudes/satisfaction'. The context was a theme park.

Whatever topic you finally select, you need to bear the following points in mind:

- Your project should make a practical contribution that is important in the real world; and
- It should make a contribution to literature (King et al., 1994).

So, in summary, your project should have value in the real world and make some kind of intellectual contribution.

Refining your topic

Refining really means narrowing down, or focusing, your topic. A way to help with this, is to think of your study, and in particular the literature review chapter, as being in the shape of a funnel. When you start to consider a topic, it is at this point in time, you have the full width of the top of the funnel, but when you start the actual primary research for your topic, thinking of the analogy of the funnel, it should be as narrow as the part from which the liquid runs out. Between the two stages you have focused your topic down, starting initially with a large range of literature, gradually narrowing down, rejecting irrelevant literature, as well as discussing relevant material, to reach a narrow topic that is achievable within your time and resource constraints, but worthwhile researching in terms of its intellectual and practical applications.

A good way to do this is to make use of the Brotherton's (2008) two terms: concept and context. It is very likely that both will have to be refined and focused down to create a manageable project, but if you ask yourself questions about the concept and the context, this should help with narrowing down process. If we take the example of the student referred to above who researched the theme park, the concept in her research is 'visitor attitudes'. So, what type of questions could be asked of visitors in relation to their attitudes to help give greater focus? Here are some possible examples:

- How much did the visit cost and did you think it was good value?
- What was your attitude to and experience of particular theme parks rides, such as the 'white knuckle' rides?
- What was your overall satisfaction with the visit?

Other questions could be aimed at particular visitors such as:

- How well did the park cater for children of a particular age?

A very different set of questions could be created around the idea of safety at the theme park. Accidents are not that common at theme parks, but when they happen, they usually receive much media attention. This 'question creating' could then lead to, for example, a focusing down of the concept on 'children, in a specific age group and how well do visitors believe the theme park caters for children's demands for exciting rides, but also the concerns of parents and the requirements of laws relating to safety'. This is clearly much more specific than 'visitor attitudes to….'. It is also potentially more interesting and certainly topical (which means it is new and therefore it will be easier to argue that you are researching something original). By focusing on the 'pleasure side' (excitement/thrills) and also the 'safety angle', you are researching in a relatively controversial area, which should mean your work is interesting (and hence maintain your attention during the process of working on your study) and the findings will have potentially a practical as well as intellectual contribution.

However, there is also the other important term to consider: *context*. If you wanted to conduct the type of research with a focus on children's rides, you would need to find a specific theme park which has such rides, but also would allow you to conduct this type of research. You would probably want this to be very accessible to you in terms of its distance from you and the time taken to get there. So 'any old theme park' will not do, it will need to be a specific park with what you require to conduct the research. Hence, this will be achieved by focusing down the context for research.

One more example is used below to indicate the nature of this focusing down process. A possible research study topic, 'quality in bars/restaurants/pubs', has been referred to above. In this example, the concept is 'quality', the context is 'bars/restaurants/pubs'. Focusing down on 'quality' could involve a consideration of what precisely quality means here. Would it be food quality? Would it be drink quality? Could it be something like 'atmosphere/ambience'? Or could the focus be on the quality of service? Moving to 'context', should the focus be on public houses, or restaurants, or bars? Even within one of these groups, there are many differences between for example, restaurants with externally recognised rating levels, those with different cuisine, those that are part of hotels and those separate from hotels. If you visited restaurants at lunch time, you would probably find different customers than in the evening and also there are likely to be different customers at the weekend compared with weekdays. Hence, the context is very important for your research in this example. Your focusing down could lead to a provisional topic 'Service quality in three selected Indian restaurants in the area Y, of City X', which is much narrower than the original 'quality in bars/restaurants/pubs'.

Although it is important to focus down with regard to both the concept and the context, there has to be a balancing act, particularly in relation to the context. If you have decided that you want to focus on one organisation (this could be the one you work for, but please read Case Study 2.2 on this topic, before selecting this location) as your context, then you have to ensure that you can get access to all the information that will enable you to complete your dissertation. You should not rely on just one or two key players to be your respondents and major sources of information. They may be keen to help at the beginning of the research, but the nature of some of your questions may make them uncomfortable and mean they withdraw, before you have gained any useful data!

To avoid the issues raised in the preceding discussion, it can be better to have more than one organisation involved. Let's assume that you have selected two organisations to research. If things go well, then you will be able to compare results from the two different organisations, which should greatly assist in the analysis process. If things do not work out well in one organisation, you will at least have the other one to provide results that you can discuss. However, you also need to balance issues to do with obtaining enough data, with issues concerned with having the time available to conduct the type of research you believe will be necessary. So, you will need to think very carefully about the scope and scale off your research. The balancing act you will be involved in means considering if the scale is 'big enough' to allow useful data to be collected, but 'small enough' to allow you to have the time to collect the data, and in terms of scope, enough variability in the process to enable useful comparisons to be made and hence offer opportunities for worthwhile analysis. This process, like most of that within the research study, will involve trade-offs and usually means limits on both the scope and scale of the research.

Research in the workplace

Some students may be employed, working outside their university course, and can see opportunities in the workplace on which to focus their research project. However, be warned this approach can be much more difficult than initially realised and it may present significant problems. Case Study 2.2 indicates some of the issues that could arise if you decide to use your workplace as the basis for your study.

Case Study 2.2: Should I research my place of work?

Many students have part-time work in bars/public houses/cafes/restaurants/fast food outlets/visitor attractions. Let us assume you are such a student. You realise that you are in a very good situation to conduct research about tourism, leisure or hospitality. This is because although your work is part-time, you work closely with other staff who are full-time. You decide that asking other staff their attitude to their work will be the focus of your research. The provisional title is 'Staff attitudes to working practices at bar/public house/café/restaurant/fast food outlet/visitor attractions ….X'. You believe this to be an excellent topic for the following reasons:

- You know there is a large amount of literature on working practices in hospitality and tourism, so there should be no problem with secondary sources.
- You believe you will definitely be able to gain access and select a sample of respondents.
- You should be able to gain a good deal of relevant information from primary research with the staff.
- Your results should be detailed and in-depth.
- You will be able to analyse your primary results in relation to the extensive literature on the topic.
- Your results are likely to have practical applications.
- Your results could lead to recommendations to change, or even improve, working practices.

However, there are potentially some major issues to be dealt with, shown below as questions, that you should consider before you proceed:

- How will managers react to your questions about working practices?
- How will you ensure the anonymity of staff, particularly if they are critical of working practices or wage levels?
- How will you ensure that staff are being honest in their responses?
- How will you be able to question your boss?
- What will you do if your results are critical of the organisation where you work?
- How will you present your results to managers and staff if this is requested?
- How will your relationship with fellow staff be affected by your research?
- Will your research work threaten your employment?

Creating aims and objectives

Once you have focused down on your selected topic, it should be relatively straightforward to come up with the questions that will give even greater focus to the research. These questions may already have occurred to you as you have been narrowing down your title focus.

If we take the example of the research referred above concerned with a theme park. The focus had been narrowed down to a concern with: 'children, in a specific age group and how well do visitors believe the theme park caters for children's demands for exciting rides, but also the concerns of parents and the requirements of laws relating to safety'. Specific question would be concerned with specific types of rides, the views of adults concerning children and safety issues.

It is conventional to modify the initial questions into aims and objectives. In fact, it will be difficult to conduct your research without clear aims and objectives. The reason for this is that it is not always easy to pose specific questions that produce answers that are useful in your study. If we take the example discussed above of 'service quality in the Indian restaurants', we could have the question: 'What is service quality like in Indian restaurants?' Following research in the restaurants, this question could be answered in each, in summary, with one of the following terms: very good, good, average or poor. However, this tells us little that is new. We could just as easily study three Chinese restaurants with the same question – substituting Chinese for Indian, and the range of answers could be the same.

However, if we convert the initial question, we can create a statement that could provide much more appropriate and useful information, and this becomes your study aim. So, your aim could be written as follows: 'To investigate what factors influence service quality in Indian restaurants'. Now this statement should produce an answer which includes a number of possible factors that influence service quality. It may also tell you which factors are more important and which less important. These factors can be compared with literature about service quality and particularly service quality in restaurants. Additionally, it may reveal something about service in Indian restaurants that is not well known. This 'new' finding may have practical applications in other Indian restaurants, or any other type of restaurant, and may also be a (small) contribution to the understanding of the concept of 'service quality' – this will be the intellectual contribution, or what is known as the contribution to literature.

Aims

Therefore, an aim is a statement which gives purpose and direction to the research process and has an output at the end. If you think of the aim in terms of it having a physical presence, then it can be considered as a target or a goal. The aim is usually accompanied by objectives. We explore in more detail below precisely what an aim and objectives are. But having created an aim and objectives means you can then look back, after you have carried out your primary research and ask yourself the questions: 'Have I met my aim, have I met my objectives?'

Objectives

An aim is usually a general statement that encompasses the whole of your selected topic. An objective is, therefore, smaller than an aim and will involve only a part, or component, of what you are trying to achieve with your aim. However, an objective is something that has to be achieved or completed to ensure that you reach your aim. It may be that your objectives are set out in a chronological sequence. It could be that you use one or more objectives to indicate a specific research technique to achieve your aim. It may be that the first objective relates to the use of a pilot study that enables subsequent stages (indicated in subsequent objectives) to be achieved.

Almost always aims and objectives begin with the word 'To' and the word is usually followed by an action word – a specific verb such as 'investigate' 'evaluate', 'determine', 'analyse'. This means, at the end of the research, you will be able to make a statement saying whether you have been able to confirm your aim and objectives or not, whatever the case may be.

Hence, the objectives are related to, but not the same as the aim, in that they are mainly concerned with *how* the research is to proceed. Objectives are often arranged in chronological sequence. Usually, the aims and objectives will be located in the introduction chapter. There will also be reference to the aims and objectives in the Conclusions chapter – the last in your study. This will have a discussion of whether the aims and objectives have been achieved and the consequences of this.

Case Study 2.3 gives examples of aims and objectives for three different types of student research project. Although these examples each have a very different focus, with one involving an investigation, and another developing a model that is then evaluated, note that each aim begins with 'To', and the objectives are very similar. The objectives in the three cases are concerned with a review of the literature, which is followed by some primary research, then analysis of this research, then drawing conclusions from the analysed findings and, finally, making some kind of recommendations on the basis of the findings of each research project.

Case Study 2.3: Aims and objectives

Example 1: Title: Visitors' perception of interpretation at World Heritage Sites

Aim

- To investigate the perception of the effectiveness of heritage interpretation at World Heritage Sites (WHS)

Objectives

- To review literature on heritage interpretation
- To review literature on interpretation at WHS
- To conduct questionnaire research using a sample of visitors to three WHS
- To interview selected staff involved in the management of the three heritage sites
- To analyse the results from both the questionnaire and interviews, draw appropriate conclusions and make any recommendations about WHS interpretation.

Example 2: Title: Destination branding: developing and evaluating a new model

Aim

- To develop a new model of destination branding and evaluate this model

Objectives

- To review literature on destination branding
- To develop a model of destination branding using literature
- To evaluate this model using a panel of experts
- To make appropriate conclusions from the findings of the evaluation of the model
- To make appropriate recommendations based on the analysed findings from the evaluation of the model.

Example 3: Title: Service quality in four-star restaurants

Aim

- To investigate customer views on service quality in four star restaurants

Objectives

- To review literature on service quality
- To review literature on service quality in hospitality and specifically restaurants

- To conduct questionnaire research using a sample of customers at the restaurants
- To interview selected waiting staff working at the restaurants
- To interview selected staff involved in the management of the restaurants
- To analyse the results from both the questionnaire and interviews, draw appropriate conclusions and make any recommendations about service quality in the restaurants.

Finally in this section, to assist you in gaining a clearer understanding of the nature of the literature review, Figure 2.1 provides a summary of what should and what should not be included, in the review.

Figure 2.1: What to include and what not to include in the literature review

What **should be** in the literature review? Summary

- An *actual discussion* of relevant ideas.
- A discussion based on a variety of sources mostly academic (journals, books, book chapters, internet) popular sources (newspapers, magazines, internet) other sources, (documents, reports, photographs).
- An argument leading to a conclusion in the appropriate section.
- The context in which your dissertation is located.
- Proper references to the works you have used.

What **should not be** in the literature review? Summary

- All literature on the topic
- Merely a list of relevant material
- A simple presentation of relevant ideas
- Material only sourced from the internet
- Material plagiarised from different sources
- Only your ideas on the topic

An important question as you prepare your study is: Where does the Literature Review fit? If you have used a quantitative research approach (see Chapter 3 and Chapter 4 for further discussion of what this means) then it is probably best located near the beginning of the dissertation. In a quantitative study, the literature review is the first major chapter in the research project, following the 'Introduction' chapter. However, on occasions students may spread this review over two or three chapters. Nevertheless, with a maximum word limit for most dissertations, (although thesis will usually be longer) extending the literature review beyond one chapter risks the problem of having to reduce the number of words in other chapters, which arguably are more important as they set out your methodology, your results and the analysis of the results.

You should be aware that the 'Methodology' chapter also have a discussion of literature, as you will refer to research philosophy and techniques you have used and you will need to defend your selection of philosophy and techniques. The discussion of literature in your 'Methodology' should be written in a similar way to your topic-based 'Literature Review' chapter.

Writing this chapter in your study

An important point to be aware of is that a literature review is not a literature survey. A literature survey is likely to be the starting point that leads to the review. However, a survey is primarily an inspection and assessment of the literature relevant to your field of study. During the survey you ascertain which written material will be relevant. More specifically, the survey should reveal which material is more relevant and which is less relevant, and which is not relevant at all. It is very likely that most of what you initially read is not specifically relevant to your study. However, as you continue to read you will find material, which is often that referred to by the author(s) in the material that you are currently reading, that is particularly relevant for your study. So, the literature review is a written account of the literature that is relevant to your chosen topic. But also remember that the literature review needs to indicate to the reader that you know more than you are actually going to write about in detail. In other words, you need to show that you know where your research fits in the wider literature, even if you do not discuss this wider literature in any detail.

The literature review, however, is not a list of the material. It is a discussion in a critical form of the relevant material. It is called a review because you are evaluating and criticising the material being discussed. You are writing to explain what the authors have written, to *critically evaluate* it, and to compare this with other related work. This means you are presenting a discussion in which you should try to write in the type of format presented in Case Study 2.4. Note that no specific topic has been identified in this example, as the point it is to indicate the generic writing style. The case study gives an indication of how a typical literature review can be written.

Case Study 2.4: The style of the literature review

'In relation to...(*your topic/issue*)..., based on results of a face-to-face questionnaire survey, Jones (1996) argued that...*(New sentence)* Smith (1998) made similar claims, and although she also used a questionnaire survey, the context of her research, in relation to the location and the sample involved, was different. Smith also indicated.........was a major finding. Nevertheless, Smith agreed with Jones when discussing the importance of the issue of......... (*New sentence*) Patel (2005), employing an on-line survey made use of both Smith's and Jones' overall methodological approach, but his focus was more on the theme of..... and the research was conducted at a different time of year. Although there was a number of similar findings to Smith and Jones, Patel indicated the issue of ...... was particularly significant.

(New paragraph) Robinson (2008) adopted a very different research strategy when exploring *(your topic/issue)*. This qualitative approach involved in-depth interviews and Robinson concluded that ...... and ...... were key factors in explaining the views of respondents. Robinson (2008) also suggested that her use of open-ended interview questions was the most probable reason for her results being different from the majority of other studies, which had used a quantitative approach.

(New paragraph) Therefore, in summary, a number of quantitative studies (see Jones 1996; Smith, 1998 and Patel, 2005) have indicated that .... and...,... are key factors, whilst an important piece of qualitative research (Robinson, 2008) indicated that......and ...... are the major themes in relation to (your topic/issue)'.

There are a number of key features in the example of a literature review in Case Study 2.4. These are as follows:

- There are a number of pieces of referenced literature;
- The similarities between the pieces of literature are indicated;
- Differences between the pieces of referenced literature are indicated;
- An indication of why there are differences in the referenced material is provided: e.g. a different context in terms of time or location; a difference in the sample of respondents; a difference in methodological approach;
- A critical evaluation is evident – the comparison of the quantitative and qualitative results indicating that each approach has produced differences in results;

- In the early parts of the example, a foundation is laid and the latter parts of the review build on this;
- A summary is provided (and this will precede a 'moving on' in the literature review).

The material in Case Study 2.4 and the bullet points that follow place the emphasis on a *critical evaluation* of literature and in relation to your aim(s) and more specifically the objectives, the term 'critically' could be inserted, in each of the three examples in Case Study 2.3 to state: 'to critically investigate the literature…'

Having outlined the key features of the review, it is important to be aware that there is no set format for a literature review. Also, the nature of the review will vary according to whether it is a deductive or inductive piece of research that forms the focus of your study. If we assume it is deductive, there are certain aspects that will probably feature in your literature review.

First, it will have an introduction which indicates what the chapter will discuss and, in particular, the major topics and issues and probably the sequence in which topics appear in the review. Second, it will probably present the major ideas from literature in relation to your chosen topic. It may present these in the chronological order in which ideas were developed and articles written. It could also present these thematically, or by a discussion of the key issues in the field you are researching. It could be arranged around the ideas of the key authors and researchers in the field. Some of this discussion will indicate the research techniques used to collect data, particularly when it is similar to the technique you intend to use in your primary research. However, the review may also focus on inconsistencies or gaps, particularly if you are proposing to try to fill in the gap in the literature through your primary research. It could be that you are suggesting that the findings from the predominant methodology and related techniques can be added to, or even improved upon, by using an alternative methodology and techniques. Third, it is likely to start off with a fairly broad approach to the selected topic, but then focus down to the precise topic you intend to research using primary approaches and then will indicate how you intend to do this. This 'funnelling down' is, quite possibly, the very last thing you include in your literature review.

The structure can be summarised as follows:

1. Start with a summary of the nature of your study and its focus (this is a lead into the major literature in the field).
2. Indicate the key topics/themes.

3 Indicate where your research will fit into the current literature (by literature is meant all that has been 'written' in any media – so could be film/TV/radio/internet as well as academic literature such as journal articles, books and book chapters).

4 Discuss the relevant literature.

Two examples are outlined below to show how this structure applies to specific dissertation topics. If your title was 'Visitor Motivation for attendance at the ---------------- Festival', your literature review would cover (at least) these topics: 1) Visitor motivation, and 2) Festival/events. Given that you have selected a specific festival, the literature review would need to cover the type of festival the visitors are attending. Let us say it is a music festival, then the specific type of festival is a music event. So, in summary the literature review should cover the following literature:

- Visitor motivation for attendance at tourism related activities
- The nature of festivals/events
- The nature of music events
- Visitor motivation for attendance at music events.

If your research topic was 'The attitude of local people to ....(a form of tourism development in the local area such as a hotel, restaurant, theme park, recreation/leisure centre)', your literature review would need to cover:

- Attitudes to tourism activities
- Attitudes of a local people/host population to tourism
- The nature of the specific tourism development/activity
- The spatial and temporal context for this tourism activity.

Finally, in terms of guidance on how to write this important scene setting chapter, Figure 2.2 summarises the processes of reading and taking notes through to the actual writing of the literature review.

Figure 2.2: A summary of the process of reading, through to writing the literature review.

1 Read the literature first

2 Think about it/reflect on it

3 Write down the ideas in your own words (DO NOT COPY)

4 Read more, repeat 2) and 3) above

5 Re-draft what you have written

Ensure that you make it flow....e.g.'X states that..... Y tends to agree, except indicates that ....., however Z has very different ideas, in particular.......'

Guide to referencing

This chapter has made several comments about the need for accurate referencing of material used in your Literature Review (and these comments also apply to any work that you have actually referred to at any point in your study, not just in the Literature Review, but also for example the Introduction or Methodology chapters). This final section provides discussion of why accurate referencing is essential and how to reference correctly.

Accurate referencing is essential in your research project for these reasons:

- You need to indicate where your research is located in relation to literature in your field and hence you should make accurate reference to the literature you have used. This should give added authority to what you write.
- Readers (particularly your supervisor and examiner) can follow up on the ideas that you have written about, by going back to the original sources. This can also enable readers to check on whether you have attributed the source of information to the correct reference.
- Avoiding actual plagiarism and preventing accusations of it are essential when writing your study. Plagiarism is when you claim someone else's work as your own. It is a serious academic offence and, as indicated earlier in this book, can lead to you being asked to re-submit your research project, or to write another! You must acknowledge all your sources.

There are a number of different styles of referencing in use. Recently in an attempt to ensure greater consistency, in many institutions the Harvard System of referencing has been adopted for those working in the social sciences. However, there is variation within the Harvard System and some authors do not follow this anyway. Consequently, there are many guides to referencing. Before choosing a system, you should check the one in use at your institution. General guidance on how to refer to different sources of literature in your study chapters and also how the references should appear in your list of references is provided below.

References in your text

These are guidelines for writing references in the text of a research project.

- In your summary of an author's work in a sentence you have written: put their name, followed by date of publication in brackets - e.g. 'Patel (1999) argues that...'. To provide variety and avoid repetition, you can also indicate in your summary the author's name at the end of the sentence, e.g. 'A research project should have a clear focus (Brotherton, 2008)'

- If there are two authors of one work (e.g. a book or article), put their names as they appear in the article, followed by the date, e.g. 'Wilson and Abercrombie (2005).'
- If there are more than two authors of the article/book/book chapter: name the first and then write 'et al' and the date e.g. 'Grabowski et al (2001)....' However, note in the 'List of References' at the end of your dissertation all the authors' names are to be listed.
- If referring to more than one work by the same author published at different dates but making the same point/using a similar argument, put in the date order, e.g. 'Hall (2005; 2008) claims that ....'
- If using a short quotation, usually one that is less than two lines, then embed it in the text, e.g. "As Clarke et al (1998: 54) argue 'attitudes are developed over time and affect the way we view the world and how we behave'."
- Longer quotations (more than two lines) should be single-spaced and indented in a separate block of text, e.g.

  > Pragmatism is not committed to any one system of philosophy and reality. This applies to mixed methods research in that inquirers draw liberally from both qualitative and quantitative assumptions. (Creswell, 2009: 10)

- If two authors have the same name then, and only then, use authors' initials, e.g. 'Beard, J and Beard, W (1994) indicate that....'.

References in the 'List of References' at the end of your study

The following are guidelines for writing reference in the List of References at the end of your dissertation.

- **Books** – put the details in this sequence – author(s) name(s), initials, date of publication, title (in italics), place of publication, publisher:

  e.g. Jankowicz, A (1991) *Business Research Projects for Students*, London, Chapman and Hall

- **Journal articles** – put the details in this sequence – author's name, initials, title, journal (in italics) volume, (issue): page numbers:

  e.g. Becker, H (1958) Problems of inference and proof in participant observation, *American Sociological Review* 23 (6): 652-660

- **Chapters in books** – put the details in this sequence – author's name, initials, title, 'in' name(s) of the editor(s) (ed. or eds.) book title (in italics), place of publication, publisher, page numbers:

  e.g. Kuo, I and Mason, P (2013) 'Managing Megalithic Monuments' in Garrod, B and Fyall, A (eds.) *Contemporary Cases in Heritage*, Oxford, Goodfellow Publishers, 130-174

- As much material is now found on the internet, you should follow the approach used in printed material, in which you give your readers the information for them to find the material. Specify the website but keep the Uniform Resource Locator (URL) as brief as possible as URLs can be very long! As websites may change, you should also give the date you accessed the site, e.g. 'http://VisitBritian.com/en/gb/ accessed April 11, 2014 or, in more detail, a complete on line reference:

  AECO (2017) Arctic Expedition Cruise Operators, on-line at www.aeco.no/about AECO/ accessed November, 24, 2019

To help with referencing, there are now several software packages, such as for example, 'Endnote', that help store references and make these available to your dissertation.

Finally, you should remember, all works cited in the text should appear in the list of references and all works in the list of references should be cited in the text!

Student activities

1 How is a literature survey different from a literature review?

2 Why do you believe it is important to take notes when reading for your literature review?

3 Select a possible topic for your study. Provide a rationale for your choice. Conduct a process of refining the topic and create a title and aims and objectives. Record how you went about this and any difficulties you had. Repeat these tasks with a different topic.

4 Looking at Case Study 2.2, what are the main advantages and disadvantages of using your workplace as the location of your primary research? Assuming you have selected your part-time job workplace, which is a fast-food restaurant, as the location for your research, write a paragraph intended for your supervisor to explain your decision.

5 What do you understand by terms *concept* and *context* in relation to a study? Using the library/learning resources centre in your institution, read the introductory section of at least two student research studies and use the terms concept and context to categorise material in these studies.

6 Using the library/learning resources centre in your institution/online resources, read the introductory section of at least two research studies at the equivalent level to yours and study the aims and objectives or research questions presented and consider how well these are related to the title of the study.

7 Explain why accurate referencing is essential in any study and what could happen if you do not accurately reference your work.

3 Research Philosophy and Research Design

Introduction

In the introductory chapter, developing self-awareness was a key process outlined and it was stated that it is possible you have assumed that the way you view the world is the same as the way that everybody else views the world. The term 'common sense' was used in this discussion. It was stated there you could believe it is common sense that the way you look at the world is precisely the same way that others look at it. However, it was also discussed earlier that one person's common sense is not necessarily the same as another's. If we accept that there are likely to be differences between different people's view of the world, it may not come as a surprise that the way some researchers view the world, is very different from others' views.

Research philosophies

The idea that there are different views of the world, and the processes that operate within the world, is part of what is known as philosophy. Philosophy is concerned with views about how the world works and, as an academic subject, focuses primarily, on reality, knowledge and existence. Our individual view of the world is closely linked to what we perceive as reality. On a day-to-day basis outside of your academic work, it would be unusual to think frequently about the way you perceive reality and the world around you. However, in relation to your study, it is very important to realise how you perceive reality, as your perception of reality affects how you gain knowledge of the world, and how you act within it. This means that your perception of reality, and how you gain knowledge, will affect the way in which you conduct the research in your study.

Qualitative and quantitative paradigms

The key term relating to the way of looking at the world is 'paradigm'. The researcher Kuhn introduced the concept of the existence of different paradigms (see Kuhn, 1970). The major reason this concept is important is that the paradigm we use to view the world, on a day-to-day basis, is very likely to influence how we conduct research. Attempting to summarise Kuhn' ideas on paradigms, Long (2007: 196) stated:

> 'a paradigm is a pre-requisite of perception itself – what you see depends on what you look at, your previous visual/conceptual experience (the way you have been taught to think) and how you look.'

As a way to start to think about what comprises the concept of a paradigm, consider the following brief example. In the early years of the 21st century, a major world bank ran a series of adverts about how it was important, when 'doing business', to understand cultural differences in a number of countries. One set of adverts had the meaning of what at first glance, appeared to be the same word in several different languages, to indicate cultural differences, while another set of adverts had a photograph of an insect, a relatively large cricket or grasshopper. Three of the different ways of seeing/viewing this cricket were as follows:

- A pest, (in parts of the USA, crickets are regarded as garden pests)
- A food item (crickets are eaten as snacks in some countries including Mexico)
- A pet (the Chinese and Japanese have kept them as pets for centuries).

So, the way you 'see' a cricket can vary greatly from country to country, culture to culture and will also depend on your individual world view.

Although each individual has a different view of the world to other individuals, it is generally accepted that there are not an infinite number of different views. In relation to research, it has become clear over the past one hundred years or so, that there are two major ways of 'looking at the world'. One view regards the world as largely objective (there is only one truth or a limited number of universal truths) and this world is measurable in terms of the use of numbers. The other view suggests that the world is largely subjective (open to several interpretations) and numeric measurement is not always possible, or desirable and hence it is words that are able to indicate nuances more accurately. In summary, these two paradigms are usually referred to as a) the *quantitative* (the world is measurable using numbers) and b) the *qualitative* (the world is subjective and is best assessed using words) (Mason et al., 2021)

Eldabi, Irani, Paul and Love (2002) suggest that research philosophies can be grouped into two general categories, which although they are the same as those stated above, use different terms: positivism and interpretivism. Hence, the links to the differentiation above are as follows: *positivism* is usually equated with quantitative research and *interpretivism* with qualitative research. As Eldabi et al (2002) claim each philosophy represents different, if not competing, views on how to conduct research. What is particularly important in terms of you setting out on your research is the point made by Phillimore and Goodson (2004) that any research philosophy adopted in a study should fit the research problem that the researcher intends to investigate, implying researchers need to choose one or other research category.

However, we will see in a later chapter and discussed in detail, there is yet a third way to conduct research, which uses a combination of the two philosophies discussed above and perhaps not surprisingly is called mixed methods research.

It is important to be aware that the research philosophy of positivism, until fairly recently regarded as the dominant approach in tourism research (Khoo-Lattimore et al., 2019), views reality as singular and independent of the researcher (Creswell, 2018; Lincoln and Guba, 1985). It also regards reality as something which can be measured objectively, for example, by using a quantitative research instrument (Teddlie and Tashakkori, 2009). However, as noted above, the research philosophy of interpretivism rejects the idea of a single objectively measurable reality, positing instead the existence of subjective, multiple constructed realities (Creswell, 2018; Lincoln and Guba, 1985), and individuals involved in this type of research, including the researcher, those being investigated, and the reader or audience interpreting the research, are regarded as the creators of these realities (Creswell, 2018). Interpretivists contend that multiple realities can only be captured through the employment of less rigid data collection techniques than generally used in quantitative research (Creswell and Plano Clark, 2007).

The research philosophies of positivism and interpretivism also differ in their assumptions about the relationship between reality and the researcher (Creswell, 2018). For interpretivists, it is argued that to understand peoples' behaviour, researchers need to put themselves in the place of the enquiry subjects (Jennings, 2011). Therefore, interpretivist researchers try to minimise the distance between themselves and respondents (Creswell, 2018). So, the interpretivist tries to get inside the respondents' minds and see the world from their view (Jennings, 2011). Interpretivists have been described as observers 'from the inside', the term used is '*emic*' (Teddlie and Tashakkori,

2009; Phillimore and Goodson, 2004). However, positivist researchers try to provide an independent, outsider's account of the research process, known as an '*etic*' approach (Teddlie and Tashakkori, 2009; Phillimore and Goodson, 2004) to ensure that the researcher's biases are excluded (Creswell, 2018).

When comparing paradigms there are three important questions:

- What is real (ontology)?
- How can we know anything (epistemology)?
- What methods should we use to conduct research (methodology)?

Ontology

The question, 'What is real?' is concerned with the concept of ontology, and in relation to this there are two possible responses, depending on the specific paradigm. In one paradigm, the response to the question: 'Is there a single objective truth/a knowable reality affected by a consistent set of laws?' would be 'Yes'. From the perspective of the other paradigm, the answer to the question is 'No' as everything is relative, there is no such thing as one objective truth or even universal truths, but merely a number of subjective truths.

Those who believe there is a single objective truth are usually referred to as '*positivists*' (there is more discussion of this term below). Such people believe there are universal truths that are waiting to be discovered. While those who believe there is no reality other than what individuals create in their heads are known as '*constructivists*' (or '*interpretivists*'). The term constructivist has emerged as those who use this approach and who believe, in relation to research, that there is no objective reality, but that reality is constructed by each individual. Therefore, to such people reality is subjective. *Phenomenology* is the term given to the research approach of such researchers (there is more discussion of phenomenology below).

Epistemology

The response to the question 'What is real?' affects the way in which knowledge is obtained. So, following on from the question 'What is real?' is 'How do we know anything about the world?' What we perceive of as reality has an effect on our knowledge of the world. Hence, each of the two different paradigms not only has a different perception of reality, but a different perception of knowledge about the world. In other words, what we think of as real, affects the way we gain knowledge.

If we perceive the world as having a number of universal truths, then these truths can be 'discovered' by carrying out 'objective' research, in which the researcher does not interact with what is being researched. In this context, neutral, objective research will be the appropriate way to gain unbiased knowledge. However, if we see the world as having multiple, contextualised 'realities', rather than objective, universal truths, then an appropriate way to gain knowledge would be for the researcher to interact with those being studied, in an attempt to reveal their attitudes and behaviour in relation to whatever is being studied. In summary, the way we perceive reality influences how we believe knowledge is gained and the process of obtaining that knowledge as a researcher. The key new concept here is '*epistemology*' which is concerned with how we gain knowledge.

Methodology

If we accept that our understanding of reality affects the way we gain knowledge about it, then we need to accept that this will have an impact on how we actually conduct research about reality (or as this is termed 'methodology').

The links between the important concepts of ontology, epistemology and methodology are neatly summarised by Taylor and Edgar (1999:27):

> 'the belief about the nature of the world (ontology) adopted by an enquirer will affect their belief about the nature of knowledge in that world (epistemology) which in turn will influence the enquirer's belief as to how that knowledge can be uncovered (methodology).'

Teddlie and Tashakkori (2009) also provided a useful summary of the contrast between each of the two conventional paradigms. When discussing epistemology, Teddlie and Tashakkori indicated that, in terms of the relationship between 'the knower and the known' (in other words the researcher and what the researcher is researching), in the quantitative approach, the researcher and what is being researched are viewed as independent of each other, whereas in the qualitative approach, they are interactive and inseparable. In terms of ontology, Teddlie and Tashkkorri (2009) state, quantitative researchers believe that reality is single and tangible, whereas qualitative researchers view reality as constructed and hence multiple. These differences in ontology and epistemology mean that different research methods have been employed, with quantitative researchers using *deductive* approaches, whereas, in contrast, qualitative researchers have tended to use *inductive* approaches (Teddlie and Tashkkorri (2009).

Deductive research

The previous paragraph has introduced two new concepts – deductive research and inductive research. Deductive research, which is often used by quantitative researchers, involves the application of known 'laws' or previous theories. Those engaged in deductive research will often apply a well-known theory to a new context (such as a different location, at a different time of year or with different respondents). Because research has been conducted before on this topic, and there are significant amounts of literature about the topic, the researcher can predict what will be revealed (this is the process of deduction). This means they can use the earlier research and argue 'because this happened before it is very likely to happen again'. Some researchers will go further and create a formal hypothesis. A *hypothesis* is a testable statement (Clarke et al., 1998). A researcher puts forward the hypothesis and then collects evidence in the research process to test (prove or disprove) the hypothesis. There is more discussion of the nature of hypotheses in Chapter 4.

Deductive research is the type of research that most natural scientists use. As indicated above, the deductive researcher starts from what is known – usually theories or even laws. The key process will involve testing the theory or applying the law. The rationale for conducting the specific new deductive research may be a different context (e.g. location or time) and although this may be different, the researcher makes an assumption that the theory will apply, whatever the context. The starting point for this type of research will be a review of the literature to determine what theories apply and can be tested. The literature review will be followed by the creation of a theoretical framework to guide what follows in the research process. This may involve the use of hypotheses that can be tested and is likely to lead to a decision on the overall research design, type of research technique to be used and how data will be analysed. Whatever the context, it is usually the case that deductive research is highly structured and focused. At the end of the research the data will be compared with the theory, often via the original hypotheses, and the theory will either be confirmed or supported, or possibly modified if the research generates unusual results.

In summary, a typical deductive research approach, in sequence, would be:

1 Identify the problem.

2 Using literature, create a theoretical framework.

3 Write hypotheses.

4 Design the research.
5 Collect the data.
6 Analyse the data .
7 Interpret the data.
8 Implement or modify the theory, or possibly develop a new theory.

Inductive research

Inductive research is different from deductive research in that it does not start from a known theory. This means it is unlikely to involve the use of a hypothesis. Instead, the researcher, often considering a relatively new topic of study, or perhaps using a different research technique, is working from the unknown, so initially collects data. Following on from this, the researcher may then try to link the data to existing theories about the topic or begin to create a new theory. This form of research is much more open-ended than deductive research, in that it may be modified in terms of the techniques used to conduct it, or who is involved, if the initial research approach does not produce useful results.

In terms of the process of research, the starting point for inductive research is similar to that for deductive research. This is identifying the problem or posing the questions that need to be researched. But from here on, the process is different. As little is known about the topic being researched, there may be an apparent lack of specifically relevant literature, and therefore little or no obvious theory to use. So, the next stage of the research process will be to design the research strategy, followed by collecting data. Data collection will be done early in the research process, partly in an attempt to identify and reveal what is occurring, but also to give possible direction to later research on the issue. After collecting the data, it will be analysed and the results compared with known theory, or if there is no existing theory then new theory created from this piece of research.

Researchers who use an inductive approach are interested in gaining detailed and in-depth comments from participants. They will hope that the data reveals a social construct or a number of constructs of reality, different perceptions or subjective statements of a topic or an issue. Unlike the deductive approach, the data will not be used directly to confirm a theory. Inductive research can, therefore, be seen within the paradigm, as discussed earlier, which is concerned with qualitative approaches to research. As indicated above, within the qualitative research paradigm, reality is understood to be constructed by individuals, and there is no objective reality.

Inductive research is often seen as synonymous with qualitative research and generally uses different research techniques to quantitative research. To obtain in-depth and detailed comments, inductive researchers are very likely to use interviews or focus groups and/or various forms of observation. The data gained by these techniques is detailed and, using research terminology, is often described as '*thick*', meaning it is in-depth and there is a good deal of it from each respondent. Given that interviewing and observation are much more time consuming than using a questionnaire survey, the number of respondents (the term 'sample' is not always used in this type of research) in inductive research is usually relatively small.

In summary the inductive research processes, in sequence, are:

1 Identify research problem.
2 Design research.
3 Collect data.
4 Analyse data.
5 Interpret data.
6 Compare data with existing theories from literature.
7 If necessary, develop new theory.

Although deductive and inductive approaches may appear very different, they can be linked. In fact, in the above summary tables, the final point in the inductive processes list (point 7) could be the starting point for a deductive researcher. A deductive researcher could attempt to test the new theory created by the inductive research process! Indeed, in many large-scale research projects a team of researchers may use both inductive and deductive research, although this is relatively uncommon in tourism related research. However, this situation is beginning to change (see Mason et al., 2010, Mason et al., 2021).

Nevertheless, as a result of the differences in ontology and epistemology and impacts on methodology, for many academic researchers the two approaches are incompatible and such researchers will see themselves as *either* quantitative or qualitative researchers, using inductive *or* deductive approaches. This situation could pose problems for you as a relatively new researcher, in that you may wish to use, for example, a qualitative approach in relation to a particular issue, but find that most research about the issue has been conducted using quantitative techniques. You may also find your supervisor prefers one paradigm to another, and it may not be the paradigm that you want to work within.

Positivism and phenomenology

We have discussed above that research philosophies have conventionally been divided up under two headings: qualitative and quantitative. We have also seen that these two philosophies are themselves linked to the two key research paradigms. We have also seen that one of the two paradigms tends to use inductive research and the other deductive research. Two other words have been used in relation to the research paradigms, but these have not been discussed in any detail, as yet. These words are positivism and phenomenology.

Positivism

As indicated above, positivism is the belief based on the view that the 'real' world, which is made up of social phenomena, exists independently of whoever is looking at it. Hence, this world can be considered as objective and independent of the observer of that world. Those who consider themselves to be positivist researchers argue that to find out what happens in this real, independent world requires impartial, value-free, empirical, logical, scientific research (Brotherton, 2008).

Positivism is the philosophical approach of almost all natural scientists (e.g. physicists and chemists). Perhaps surprisingly, it is also the dominant approach in the study of business and management. This means that it is also the major approach to research in the field of tourism. Those who use a positivist approach to conduct research in the tourism field believe that those involved behave in a logical, rational way (Brotherton, 2008). Those who use the positivist research approach also believe that events can be explained by cause and effect 'laws'. These laws are considered to be universal and, although the context in which the laws exist may be different, the laws will still apply.

Ideally, positivists would like to conduct their research under laboratory conditions and their research would be in the form of experiments. Here it is possible to control important factors that affect results. This control is very important, as it should be possible to identify the key factors that affect results and to measure the more important of these in relation to other factors, which are less important. Positivists not only want to identify the important factors, but measure them as well, so they can be compared with less important factors. Positivists also want to compare the factors with previous research on the topic. However, it is important to be aware that the real world is usually far more complex than a laboratory. Nevertheless, the intention of those involved in positivist research will be to discover the key factors, what cause-effect relationship exists and what the effect actually is.

Trying to conduct positivist research becomes more difficult when research topics that are not easy to measure are involved. It is usually possible to find out how many visitors are attending an event, collect demographic information about these respondents, such as their gender, age, income and educational level, calculate how much they are spending and work out how they are spending their money. However, how satisfied they are with their experience of the event is likely to be far more difficult to measure. The reason is that 'satisfaction with their experience' is more subjective – but of course it will be very important for those attending the event. In addition, organisers of the event may be also very interested in the visitors' experience, so that they can replicate this experience at future events, providing it is a 'good' experience.

Testing hypotheses

Continuing with the theme of how positivist researchers operate, they usually want to obtain evidence to support whatever theory they believe will apply. The way a positivist researcher looks for confirmation of a theory is through the use of a hypothesis or perhaps several hypotheses. The hypothesis is created by the researcher examining the implication of the theory they believe applies and then applying the collected data to the theory. The hypothesis is in effect claiming, 'given the theory, this should occur in this particular piece of research'. In positivist research, a hypothesis will usually be written as a statement which can be tested, and it will be tested by collecting data that either supports it or rejects it. An example of a relatively simple, but testable statement, is:

> 'The leisure centre will have more families using the facilities on a Saturday between 10 and 12 noon than on Monday between 10 and 12 noon'.

This is based on the idea (which is actually a theoretical position – see below for more discussion on theory) that leisure centres will have more family use at weekends than on weekdays. This hypothesis could be tested easily by counting and could be verified via a simple questionnaire.

A somewhat more complex hypothesis is provided in the following discussion and was one I used in my own research. In the early 2000s, I conducted research at a week-long music festival and one factor I was interested in was the nature of visitors and the type of ticket they purchased. In particular I wanted to discover who purchased a season ticket (a week-long ticket), and also find out who visited on a daily basis, (buying a day ticket) and possible differences in each group's behaviour. The festival, an outdoor music festival, had its own campsite, and included in the price of buying a season ticket was 'free' camping for a week. The price of a day ticket to the festival did not include any type of accommodation. Therefore, I created the following hypothesis: 'Those visitors who have purchased a season ticket will be staying at the festival campsite, whilst those who have day tickets will be staying elsewhere' (note that this is actually a two-part hypothesis). I tested my hypothesis via a questionnaire – this was part of a larger research project, so there were questions on other topics as well. My results indicated that from my sample of visitors, 82% of those with season tickets responded they were staying at the festival campsite and 89% of day ticket holders indicated they were staying elsewhere (i.e. not at the festival campsite). Therefore, I argued I had empirical evidence to largely support my hypothesis.

Criticisms of positivism

I have indicated above that there are issues with positivism when trying to measure certain aspects of the real world, including subjective features, such as 'visitor experience'. However, perhaps the biggest criticism of positivism is that it is not possible to be totally objective in the real world. In other words, it is not possible to be completely impartial when human beings have values, prejudices and beliefs. Some respondents may not even be consciously aware of this, but these factors will still influence responses. There are also concerns about the way a positivist researcher wants to be in control in the attempt to achieve objectivity. The natural scientist can claim, reasonably convincingly, that their presence as researcher does not have any effect on the reaction of one chemical with another in their experiment in the laboratory. However, a positivist tourism researcher controls the respondent through the nature of the research technique, frequently a questionnaire and by the specific questions on it (Long, 2007; Clarke et al., 1998). The researcher knows that asking 100 respondents to answer 'Yes' or 'No' to a specific question should produce two (probably different) percentage figures. The researcher will then be able to claim they have an accurate response, and this result is reliable. However, the respondents have had no input into the research technique – the questionnaire – or the specific question (see Brotherton, 2008). Respondents may feel that they did not want to be asked that particular question in that way, but would have preferred to give a comment using their own words. The particular positivist research technique described above will not allow this – so it can be claimed that the result is not valid. It is not valid because it was the wrong question asked in the wrong way! Note that the concepts of validity and reliability are discussed in more detail towards the end of this chapter.

Yet another criticism of positivism is that, as it is seeking universal laws or truths, by definition, it is ignoring the context in which events occur. Let us take a relatively simple example involving an outdoor swimming pool. Based on research, the owners believe that customers will feel most comfortable when the water temperature is 28 degrees. However, it seems logical to suggest that the outdoor swimming pool, with a water temperature of 28 degrees, will have far more swimmers on the occasion that the air temperature is 35 degrees, than the occasion when the air temperature is only 15 degrees. In this example, context (here represented by air temperature) is all important in influencing the number of visitors, independent of the constant factor (the identical water temperature in each case).

One other criticism of the positivist approach, and in particular the use of questionnaire surveys, is that responses tend to be lacking in depth and detail. If you are trying to create a relatively straightforward easy to answer questionnaire, in an attempt to gain a high response rate, then closed-ended questions (such as 'Yes/No') are usually easier to analyse than open-ended questions (see further discussion on these types of question in Chapter 4). It is also usual for researchers, using such a questionnaire, to set a target for the number of responses that they feel confident will give them sufficient material. The emphasis therefore is on the quantity of responses rather than quality. However, closed-ended questions do not usually provide the 'reasons behind the responses'. A solution to this is to create more complex types of closed-ended questions, which include tables with a range of boxes to select from and tick the appropriate category and 'Likert scale' and 'bi-polar semantic differential scales' (these are discussed in more detail below). However, even these types of question do not usually provide an opportunity for respondents to explain why they responded in the way that they have.

Another criticism of questionnaires, as one type of positivist technique is that a questionnaire is merely 'a snap-shot in time and place' (see Mason and Cheyne, 2000). Other research techniques can also have this criticism directed at them, but a questionnaire made up of mainly closed-ended questions will be a very limited 'snapshot', lacking detail in, and explanation of, the responses. In an attempt to overcome the 'snapshot' problem, some researchers have used several questionnaires with the same focus and very similar questions at regular intervals, over time, but in the same place (see Getz, 1986; 1994). This is sometimes referred to as a longitudinal approach (see Getz, 1994). However, strictly speaking, a series of snapshots at different points in time is not a longitudinal approach, as this would require continual monitoring via research over a relatively long period. Continual monitoring is relatively uncommon in tourism research, and even the approach of several snapshots, over time, is still an unusual approach.

Phenomenology

Phenomenology is the name given to the other major research paradigm. In the discussion above, the terms 'interpretive' and 'constructivist' have been used, and to a great extent these words are synonymous with phenomenology, although they more accurately imply the techniques used by a phenomenologist.

As stated above, phenomenology starts from a different ontological position to that of positivism, which also means its epistemology and generally its methodology will be different to that of positivism. As has been discussed above, 'constructivists', 'interpretivists' or 'phenomenologists' (these terms are often used interchangeably) believe that reality is created by people in the way that they think, behave and interact. The world is, therefore, socially constructed and does not exist in some objective state.

Phenomenology emerged in the academic areas of sociology and social anthropology, largely as a rejection of what was the traditional positivist way to research, because it was felt positivism was not applicable in many social contexts (Clarke et al., 1998) As discussed above, positivism can provide accurate answers to questions. But positivism does not always provide opportunities for certain questions to be asked, or for questions to be used in particular ways. Such questions are designed to provide detailed, in-depth responses and therefore positivism often fails to explain fully *what* is occurring and, more importantly, *why* it is occurring in many social contexts.

Hence, when researching human interaction and behaviour, there have been those who rejected positivism because, although they accepted it produced reliable results and these were ones that could be replicated by other researchers asking the same questions, they argued that these results were not valid. In other words, the questions were either 'wrong' or even more likely, being asked in the wrong way. And of particular importance to the researchers who rejected the positivist approach, the results revealed little or nothing about 'real' world issues.

The importance of context

Phenomenologists do not search for objective truths but are interested in ways in which individuals and groups perceive the world (Clarke et al., 1998). Unlike positivism, the context is very important in phenomenology, and it is the context that significantly affects the way individuals perceive the world. So, phenomenologists do not try to eliminate the context, but build it into their research, arguing that the context is vital for understanding what is occurring.

As phenomenologists do not use the same paradigm as positivists, to a great extent, because they reject the positivist ontological position, (they do not accept that objective reality exists), the epistemology and methodology of phenomenologists is usually different from positivists.

Phenomenologists argue that, as humans do not usually live under laboratory conditions, they do not support the use of experiments as a research approach (Brotherton, 2008). As the context is important for phenomenologists, this is not eliminated from research, but becomes an important part of it. As phenomenologists believe reality is socially constructed, they are interested in people's views of reality and how they construct it. To gain an understanding and attempt to explain what is happening, phenomenologists will compare the different 'world views' (socially constructed realities) of respondents. Unlike positivists, they also believe that their presence in researching respondents is likely to have an impact on the respondents and may change the way in which respondents behave when asked questions or when being observed.

As phenomenologists believe reality is always contextualised and open to different interpretations, they argue that the best way to understand and explain any situation, or occurrences, is by collecting data, prior to any use of theory. Therefore, phenomenologists use inductive approaches rather than carry out deductive research. Phenomenologists believe that using inductive approaches will give valid answers to their questions. The methodology of inductive research is therefore different from deductive research, in that data is collected first, and then the phenomenologist will work 'backwards' to theory.

However, it is important to be aware that although experienced phenomenologists may approach a new research situation with an open mind, they are very unlikely to have an empty mind. Given that many experienced phenomenologists will have conducted similar research to their current project, they are likely to be influenced by what they already know from experience, which means they may have one or several theories 'in their mind' when they conduct their research. Nevertheless, phenomenologists would argue that they are not proceeding from theory to data, as a positivist using deductive approaches would do, but from data, which is then compared with current theory or theories.

Commonly, phenomenologists conduct research for different purposes to positivists – they want to discover people's interpretations of the world, rather than test a theory. This means that they tend to use different research techniques to positivists. Without a theory to create or hypotheses that can be tested in relation to the theory, phenomenologists will use techniques that produce detailed and in-depth responses. So, interviews are commonly used and also focus groups. Experiments or closed-ended questionnaire surveys are generally not used by phenomenologists. As context is very important, observation is a frequently used technique for phenomenologists, and locating the research directly in reality, means a common approach is the use of case studies, which are real world examples.

As phenomenologists are often involved in researching complex real-world situations, they tend to be attempting to find patterns, connections or relationships. Hence, phenomenologists are collecting what they would call 'rich' or 'thick', data that provides a picture of the complex reality being studied. This justifies to phenomenologists their use of a qualitative approach and related techniques and, as such, they collect predominantly words and do not generally produce results that are in the form of numbers.

■ Differences in methodology

Although the discussion above has indicated that the differences in ontology and epistemology in the two paradigms leads to differences in methodology, it is important to be aware that not all positivist research is quantitative and not all phenomenological research is qualitative. For example, a quantitative piece of research using a questionnaire survey as the main technique may also use an interview to support this. Occasionally, a qualitative study may have a part which is quantitative. However, in summary, the major points to be aware of, are that positivists usually make use of deductive research approaches and conduct quantitative studies, whilst phenomenologists nearly always use inductive research and carry out qualitative studies.

As each of the different paradigms has a different ontology and epistemology and therefore different methodology, the way that results from each approach are analysed also varies considerably. For example, a closed-ended questionnaire may be relatively easily analysed using a computer programme and statistical techniques. The results of participant observation or those from focus groups are likely to be much more difficult to analyse using a statistical technique and will require a different approach. The analysis of data is discussed in detail in Chapters 7 and 8.

Case Study 3.1 provides a scenario to indicate the differences between the overall approach and specific techniques of both positivism and phenomenology.

Case Study 3.1: Differences between positivism and constructivism

John and Jenny have just been appointed as assistant managers at an old established central London four-star hotel. Both are graduates with degrees in hospitality and tourism management. They have each been working for six months and the manager of the hotel has asked them to carry out some research in relation to a problem. The hotel has received some bad reviews on 'trip advisor' type websites over the past year and most of these relate to poor service quality. The manager wants to know the exact nature of the problem, the precise causes and for John and Jenny to suggest a solution.

John and Jenny sit down to discuss how to conduct this research. John indicates that when he did his degree, he remembers that one of his lecturers had used a theory about service quality which had been adapted for use in hospitality and tourism. He argued that what they should do is find the original research, he thinks it was by someone called Parasuraman, and then look at how tourism researchers have adapted this. Then he says, he is sure that most of the research conducted on service quality will have used questionnaire surveys, so they should adapt one that has been used before, and specifically one with exclusively closed-ended questions that will be swift and easy to analyse. He adds that they should then carefully select a sample of hotel guests, to give the questionnaire to and conduct the research immediately after the guests have had a meal in the restaurant, as this will be just after they have been served by waiters. John says that then they can be certain of getting accurate answers to their questions about the nature of the problem and its causes. Following this, they can use the answers to make recommendations on a solution, and they can present these with confidence to their managers because of the valid, robust, systematic research they have conducted.

Jenny nods her head and agrees that they need to feel confident when they present their results to the manager, but adds she is not convinced that John's is a suitable research approach, arguing it may not tell them very much. She says that it is difficult to reduce the experience of service in the hotel to a simple 'yes/no' type answer. She also says that although there is definitely a 'service encounter' in the restaurant, there are also likely to be several others, such as at reception, in the bar and with room cleaning staff. She argues that those questioned should be given the opportunity to think about all types of service they receive in the hotel. She goes on to indicate that for this reason and others it would be better not to have pre-set questions, but to allow guests to give detailed comments in their own words on service quality. She argues that interviewing a small number of guests may start to reveal what aspects of service guests are unhappy with. She also says that they should ask waiters and bar staff to observe the behaviour of guests, particularly when there is a 'service encounter' and perhaps even listen to conversations amongst guests to try to pick up any negative (or even positive) comments. Finally, she says that they can compile the responses from those interviewed and get the feedback, fairly quickly, from the staff who have been observing guests and present these preliminary findings, with recommendations, which would probably include a more detailed formal research process with guests and staff, to their manager.

Case Study 3.1 indicates the advantages and disadvantages of each research approach. I will leave it up to you to decide which seems more appropriate, given the circumstances outlined in the case study.

Types of research

It is clear, in the eyes of many people, including those in the media, politicians as well academics and students, that there is 'good' research and 'bad' research. Also, from the earlier discussion you will have become familiar with scientific research and research that could be regarded as not scientific. You may also have realised there is theoretical and applied research. Research can also be conducted in the real world or under laboratory conditions. Research can be used to test knowledge that already exists or obtain new knowledge. So, there are many different types of research.

However, in terms of classifying research relevant to your study, it may be better to think of the purpose of the research (Brotherton 2008). It is conventional to subdivide categories of research by their purpose under three headings: exploratory, descriptive and explanatory.

Exploratory research

Exploratory research is used when an issue or topic is new. If a new topic has emerged in the field of study, an exploratory approach would be useful to gain some initial insights (see Mason et al., 2010; Mason et al., 2021). Alternatively, it may be that it has not been possible to conduct research into the topic before – for example a government may have restricted access to sensitive documents concerning a radio-active leak that has polluted the sea at a coastal destination, so an exploratory approach could be used in this context.

The effect of COVID-19 on tourism is currently a new topic. It could also be the case that a topic which has been researched before is exposed to change, for example, via technological advances or modifications in consumer demand. This would also be an opportunity for exploratory research. However, exploratory research is often not the only type of research in a project. It is very likely to be the first part of a project and enables a research agenda to be set (Brotherton, 2008; Mason et al., 2010). This can then be followed up by other types of research.

Descriptive and explanatory research

The types of research that could follow on from exploratory research are termed descriptive and explanatory, although it should be noted that neither of these actually require exploratory research in the initial stage. Descriptive and explanatory research are the major types of research in the tourism field of study.

Descriptive research is primarily concerned with establishing an accurate picture of a particular issue being researched. It could be that within a topic, it is known that certain events occur, but the sequence in which they happen may not be clear. Descriptive research can be used to reveal the order of these events. It can also be employed to classify categories, describe processes or relationships, or provide a profile of a particular group (Neumann, 1984).

Descriptive research does not intentionally seek to explain phenomena, but this is precisely what explanatory research is attempting to do. **Explanatory** research is trying to answer the question 'how' something occurred and of most significance 'why' it occurred. Explaining why, or attempting to explain why, is often the key part of a research project.

There is often a link between descriptive and explanatory research. For example, in the chronology of events, a piece of descriptive research establishes there is a relationship between factors, or variables in a project. The explanatory part of the research will then attempt to explain how the relationship occurs and why it occurs. Explanatory research can also provide evidence to fit a theory, test how accurate a theory is, or actually help create a new theory.

Of course, it is possible that a research project could be exploratory, descriptive and explanatory. However, in terms of your dissertation, it is probable that time and resource factors mean that you cannot be involved equally in all three types of research. Most student dissertations therefore tend to fit into the 'descriptive' or 'explanatory' categories. Many students may feel that it is necessary to conduct an explanatory piece of research. This is not necessarily so. For instance, a very good descriptive piece of research, which provides factual material on a subject that is not that well documented and therefore not fully understood in the tourism field, may well be preferable and score higher marks, than an attempt at explanatory research which does not accurately answer the questions 'how' and 'why'.

It should also be noted that the great majority of student research projects will almost inevitably contain some descriptive and some explanatory elements. So, it is not the case that the research is one type rather than another, but that the emphasis is on one type more than the other, though not to the complete exclusion of the other. Therefore, one study could be largely descriptive but with some attempt at explanation, while another may be focused on explaining a set of circumstances following on from a brief description of the nature of the issue being researched.

Primary and secondary research

There are some other important distinctions in relation to types of research that are likely to be particularly important in relation to your study. You have probably heard the terms primary research and secondary research. Secondary research makes use of already published material – it uses whatever is there in the literature on a specific topic. This type of research usually involves some kind of new analysis or re-interpretation of whatever has been published on a topic previously. It is unlike primary research in that primary research involves the collection of new material. It is customary for this new data to be then compared with what has already been published to look for similarities and differences (in other words primary data is being compared with secondary data). It is frequently, although not always, the case that you will be required to collect primary data in your study. You should check your institution's regulation on this.

Theoretical and empirical research

Another important distinction is between theoretical and empirical research. Theoretical research is most commonly secondary research. It involves looking again at secondary data, reflecting on what has already been found in the existing studies and possibly attempting to re-interpret the original findings. This is likely to involve the secondary material being questioned, re-evaluated and possibly revised and this may lead, as a result of this process, to the creation of a new theory. Empirical research always involves the collection of new data. This gathering of new data will, by definition, involve primary research – as primary research involves the collection of new data. Therefore, the terms empirical research and primary research are, to all intents and purposes, synonymous.

A useful way of remembering the differences between primary and secondary approaches is to think of primary as being 'first-hand' (new), whilst secondary is 'second-hand' or old (Brotherton 2008).

Theory and hypothesis

The discussion about the use of the deductive approach has made reference to applying theory and testing hypotheses. In reference to the inductive approach, theory has been discussed in relation to what happens after you have gathered data. So, in both deductive and inductive research, theory is important. But what exactly is theory? Cooper and Schindler (1998:47) indicate that it is a set of systematically interrelated concepts, definitions and propositions that are used to explain and predict phenomena. As Brotherton (2008) states, theory helps us understand what has happened in the past, what is happening now and what will happen in the future. It is probably this last point, that theory can be used to predict what will occur in the future, that is the most important point. This is one of the major reasons why theories are regarded as so important in the academic world.

Case Study 2.1 (see Chapter 2) discusses a particularly important theory in tourism, and this is relevant to the discussion here. The Tourism Area Life Cycle (TALC) is a theory that has emerged in tourism, albeit borrowing a part from the field of marketing! A key aspect of the TALC is that it is based on the ideas of several other important researchers and theorists working in the field of tourism and related areas. The concepts and constructs developed by these earlier theorists were used by the researcher Butler in the late 1970s in combination with the marketing theory of the product life cycle. The product life cycle indicates that there is a relationship between a product and the time over which it is being sold. Simply stated, when a new product is launched there are very few sales, but with appropriate marketing, sales increase and eventually, over time, reach a peak number of sales. After this, sales will tend to decline, unless there are attempts, via marketing, to re-launch the product.

This product life cycle theory is usually represented by a graph, with time on the horizontal axis and volume of sales on the vertical axis. The line on the graph starts with a low point in the left-hand corner (launch of the product) which then rises steeply to the right (reaching a peak of sales at some point in time), before falling way again. The curve is, therefore, a leaning forward S-shape. Butler took the graph shape and used time on the horizontal axis and substituted growth of a destination for volume of sales on the vertical axis. What was shown on the graph was a series of stages of growth of a tourism area or destination. The beginning, with emergence, passing through development, consolidation and eventually reaching a peak of development, before stagnation and possibly decline. TALC was seen to be a very useful theory, particularly if it could be ascertained at what point on the graph (meaning at what stage of development, or what point in time) a particular tourism area (or tourism destination) was at when it was being studied. If this could be worked out, then the TALC theory could be used to predict what could happen to tourism destinations in the future. Therefore, it was seen as very useful for planning purposes. The theory was also relatively easy to understand, so could be tested out on destinations around the world to discover if the theory applied, or not, in relation to specific destinations. Since its appearance in 1980, it has been applied to many destinations in the past years or so and also frequently critiqued.

In summary, the TALC has key features of many useful theories.

- First it draws upon other theories that have been developed before, which give it a certain credibility
- Second, it can be applied relatively easily to any tourism area or destination in the world.
- Third, it can be used to predict what will happen in the future.

Testing out a theory usually involves the use of hypotheses. A hypothesis is, put simply, 'a testable statement'. Examples of hypotheses have been presented earlier. The major point, in terms of what is being discussed in this section, is that they have emerged from theories and evidence can be collected, via primary research, which can be used to confirm (or reject) a hypothesis. When we are testing hypotheses, we are concerned with relationships and often, although not always, the relationship is one in which one factor causes, or is believed to cause, a specific event or activity. Therefore, a hypothesis can be used to test whether one factor causes another. However, it is customary when testing a hypothesis to making a negative statement – which is called the 'null hypothesis.' It is stated in this way due to the sceptical nature of the scientific approach (Brotherton, 2008). So, we start by making a negative statement. If we take the example above of the number of visitors to a leisure centre on different days of the week, the null hypothesis would be: 'There is no relationship between the number of leisure centre users and the day of the week.' If, (as is very likely) our research revealed far more users on Saturday during the period 10-12 than on Monday between 10-12, then we would state that we have rejected the null hypothesis and would accept that there is a relationship. Note, that we have not at this point argued that one factor causes the other. However, we could do this, arguing that it is the day of the week that causes there to be different numbers of leisure centre users. Putting this another way, the numbers of users, depends on the day of the week. We now have two variables that we are researching the relationship between. The 'number of users' is referred to as the '*dependent* variable' and the 'days of the week' as the '*independent* variable'. Therefore, what we would be investigating is whether there is a causal relationship between the dependent and independent variable, or in this case the 'number of users' and the 'day of the week', respectively.

Variables

The previous discussion has introduced the idea of variables for the first time. Discussed above are two types of variable: the dependent variable and the independent variable. There are other types of variable, as well as these two, and the following example indicates a range of variables. If we imagine an athlete who is training in an attempt to gain a medal at an upcoming championship, it is possible to refer to both independent and dependent variables. If we assume that the athlete trains every day, then it is very likely that success at the forthcoming events will depend on the type and amount of training. So, using research language, the success of the athlete (the dependent variable) depends on the amount of time and effort (the independent variable) that the athlete invests in training. We can also introduce two other variables here. Let us assume that the training carried out by the athlete leads them to be fitter and more confident in their ability and they then have the possibility to go on to gain a higher placing in the championship. This can be referred to as the *intervening* variable. However, let us also assume that as an amateur athlete, there is a requirement to work part-time to enable the athlete to have some form of income. This will obviously have some effect on the amount of time and effort that the athlete puts into training. This is known as the *moderating* variable.

In much research, it is not easy to identify the intervening and moderating variable, because of the complexity of a particular situation. Nevertheless, the independent and dependent variables are very important particularly in deductive research, as what is being investigated is the relationship between them and often that it is a 'cause-and-effect' relationship. A hypothesis, in this case, will be used to test if there is (or is not) a causal relationship. In effect, what is being tested is the effect of the independent variable on the dependent variable.

In deductive research, if hypotheses are to be created and tested, then the dependent and independent variables will need to be known in advance to create the actual hypothesis. In most research of this kind in which a hypothesis has been created, it will be the dependent variable that is the key factor being investigated. In my example, above, of research into festival tickets, the main focus of this part of my research was how ticket types could influence accommodation choice. Restating this as a research question gives: 'Does the type of ticket that a visitor purchases influence their choice of accommodation?' To create a hypothesis from this question, we need to be aware that the type of ticket is the independent variable, and the dependent variable is the choice of accommodation. Hence, the hypothesis could be written as: 'Festival ticket type affects accommodation choice'. My actual hypothesis was more specific than this and subdivided into two related parts ('Visitors who have purchased a season ticket will be staying at the festival campsite', 'whilst those who have day tickets will be staying elsewhere') but was nevertheless a 'cause and effect' type hypothesis that could be tested.

Studying a range of variables is at the core of much good research, and very often what is being researched is the relationship between dependent and independent variables. It should not require much effort to realize how many variables could be researched in tourism. However, below are some more examples of hypotheses to suggest the range and scope of these in the tourism field:

- 'The number of air passengers travelling in Europe has increased rapidly as a result of the growth of low-cost budget airlines'.
- 'As air temperatures rise in a specific coastal tourism destination, more tourists will swim in the sea'.
- 'During 'happy hour' a public house will have more drinkers, compared with one that does not have a 'happy hour' at that time'.
- 'There will be a greater demand for tickets if a low division football team plays a Premier league team at home in a cup match, than for a normal home league game'.

All these statements are hypotheses having dependent and independent variables and all could be tested by collecting data and confirming or denying the hypothesis. A similarity between all of them is that they show a positive relationship between the variables – as one goes up so does the other. This does not always have to be so. Take this statement: 'Fewer customers will order the expensive main meal menu options compared with the cheaper options'. This shows an inverse relationship – in other words as one goes up (here it is price), the other (here number of customers) goes down. Also look more carefully at the statement above about happy hours and public houses, although the emphasis is on the pub with the happy hour leading to more drinkers, if we turn this statement around, we can state: 'a lack of happy hour means fewer drinkers'. Now, both aspects are negative and there is therefore a negative relationship between the variables.

Choosing your research philosophy and specific techniques

Much of the previous discussion in the book has indicated the nature of various research philosophies and the rationale for these. However, in relation to your chosen topic in your study, there will be a point at which you will have decided how you are going to conduct your own primary research. This may be comparatively easy, if you have realised that you want to collect data that can be easily converted into numbers. Or you may be testing a theory via the use of hypotheses. It could be that you have decided that observation of a group of people involved in an activity will be the best way to gather useful information on your topic. In these cases, the overall philosophical approach may be relatively easy to decide on. However, it could be (and often is the case) that you do not really know how to proceed with your study!

One important way to help you select your philosophical approach, overall methodology and specific research techniques, is to carefully look at and reflect on your title, your aim(s) and your objectives. Probably, your aims and objectives will be most helpful in this process. You need to ask yourself questions such as the following: What am I trying to achieve in my aims and objectives? Am I exploring a new tourism topic? Am I testing a theory? Have I included reference to creating and testing hypotheses? Is there some form of possible cause and effect relationship that I am investigating?

It could be that your objectives already indicate preferred techniques to gain the data to achieve your aims. If you have specified certain techniques, these are commonly found more in one research paradigm than another. As indicated above, using observation is most common in inductive, phenomenological, qualitative research, while a questionnaire with closed-ended questions will usually be a part of positivist, deductive, quantitative research.

What is very important is that you select your approach carefully and are able to give the rationale for the choice. You will have to choose between, not only the overall philosophical approach, but also the specific techniques, and then justify this (in other words give a clear rationale for the techniques chosen). This is a very important part of your dissertation. Remember that your supervisor (and examiners) are not necessarily experts in your chosen topic, but they should be experts in research methodology and techniques. They are looking to see that you have not only selected an appropriate way to conduct your research, but that you know why this is appropriate. If you are asked why you selected a particular technique, and answer something to the effect: 'My supervisor told me it would be a good idea to use that technique', and only state this as the reason, this will not help you convince anyone that you are a good researcher!!

If you are testing a theory and using a hypothesis or several hypotheses, then you will be involved in deductive research. It is most likely that your primary research will be quantitative, and you are also very likely to use some form of questionnaire survey in your research. If, however, your research is to focus on a relatively new topic in the tourism field, it will be largely exploratory. You will be intending to gather in-depth and detailed responses from participants. In this case, your research approach will be qualitative.

When making your choice, be aware that there are no techniques that guarantee one hundred per cent success. All techniques have advantages, but also disadvantages. Also, the particular circumstances in which you find yourself, when doing your study, means that you cannot always do the research in the way that would achieve the best results. You are certainly limited by time and resources. You may believe a postal questionnaire survey sent to a sample of 20,000 possible respondents would really be the best way to conduct your research. But do you have the time and resources to send out 20,000 letters? If your response rate was 40%, how would you analyse 8,000 questionnaires? How much time would this analysis take? You will need to make a decision, in which you weigh up the specific technique, against the time and resources available. When you defend your decision, you can refer to resource and time constraints, but you still need to select an appropriate technique. In effect, what you will need to argue is that the overall approach and the techniques used are 'fit for purpose'. This means that when your supervisor and examiner read your study, they will need to be able to see and understand the rationale for your choice of approach and techniques. Your write up of this needs to be transparent, so that the reader feels confident that you have used an appropriate approach and techniques, so that they also are able to confirm that your results have credibility.

One way to gain confidence in your choice of research approach and specific techniques is to explain what you have done, or are proposing to do, in spoken words, face-to-face with a friend, another student or, if you work part-time, a work colleague. Although it may be best to do this with someone who has a reasonably good understanding of research, sometimes trying to explain what you regard as complex to a colleague, who has little idea, although this can be difficult could be very helpful to you. However, it is to be hoped that your friend, whatever knowledge of research they have, will ask questions, with 'clarification' and 'explanation' type questions being very likely. As you attempt to respond, you should get a feeling of how well you can explain, justify and defend your primary research. If you cannot do this easily, in other words if you cannot provide good arguments for your research approach, it could be that you have made some poor decisions on what to do or they were made on the wrong basis and this will not lead to a good study, and obviously not a good mark!

Major empirical data gathering approaches

As indicated above, there are potentially a large number of ways to conduct both quantitative and qualitative research. However, as there are constraints on what can be done in a student research project, not the least being limited time and also resource constraints, particularly that there will be only one person (you) conducting the research, this leads to some research approaches being viewed as preferable to others.

Probably the most common approach adopted by students (at both undergraduate and postgraduate level) is that of the questionnaire survey. The specific, detailed nature of this in terms of e.g. question creation, question type and ways to analyse will be discussed in Chapter 4, which focuses on quantitative research. However in this chapter, the attraction and nature of questionnaire survey research is outlined. In addition, as one of the major data gathering approaches used in qualitative research, the use of case studies is discussed.

■ Questionnaire survey research

It is a very unlikely scenario that you have not looked at some form of questionnaire survey already in your life. Surveys are used commonly, for a great range of reasons and by a large range of organisations. These vary from political polling organisations to marketing companies. They are given to a wide range of respondents, including business customers and employees to provide feedback, and are also a favourite tool of social science academics. Hence, questionnaire surveys are familiar to most people in wider society, but also are used in academia. It is the fact that they are well known research tools that makes them attractive to many students.

It is important to be aware that not all surveys are questionnaire surveys. It is possible for one researcher to conduct a survey of, for example, the users of facilities in a leisure centre, or the condition of five swimming pools in a city without asking anyone any questions. The researcher here is merely using a checklist, and possibly adding extra comments, but is nevertheless involved, in both cases, in survey research. Therefore, questionnaires are one type, although a very common type, of survey.

Questionnaire surveys come in a variety of formats: they can, for example, be postal, electronic (on-line) or face-to-face. Whatever their format, they have in common that a sample of the overall population will be involved. The sample will be asked questions about a specific topic or theme or related themes. The survey also involves a communication process (Brotherton,

2008). The survey producer has to make clear, via the questions, what is wanted from those being asked to complete the survey. If the communication process is not clear, then it is quite likely that the respondents will not be able to produce good answers. Outside the academic world, surveys are often carried out by, or on behalf of, commercial organizations that are trying to sell products, and in this context, if the communication process is poor, this will be a costly activity that yields little useful data. If you use a survey in your study, you are not likely to be wasting money, but it is a good idea to be aware of what could happen to you if you were working on behalf of a commercial producer and due to a poor design, your survey did not gather useful data.

Is your sample representative?

A survey uses only a sample, as using the entire population would be too costly in terms of both time and money. However, using a sample is always beset with problems. The key point is that the sample must be representative of the whole population and ensuring this is the case will be a major issue if you decide to use a survey in your project. For example, if you had the time and money, it might be possible to get the entire population of the UK to complete a survey of their holiday plans for next summer. (Here 'population' means all those who are in a position to take a summer holiday). This would mean millions of possible respondents. You are probably aware that large polling organisations assess the voting intentions of the entire eligible voting population of the UK (40+ million people) using a sample of anywhere between 1000 and 3000 people, which is very small percentage of the total UK population. How do they do this? By ensuring the sample is representative of the entire voting population, according to the demographics of this population. So, factors such as gender, age and ethnicity, as well as geographical location will be key criteria in determining the sample. Sampling is discussed in more detail in the Chapter 4, which focuses on quantitative research.

Advantages and disadvantages of surveys

One of the key advantages of surveys is that they can be quick and easy to design and then put into operation, compared with some other forms of research technique. Providing the design has been good, the results can be reliable, they can also be easy to analyse, and in many cases statistical techniques can be used with them. This means that it should be possible to compare the results of your survey with other similar work, and if this is relevant work from the literature, this will give your study greater authority. It may also give you the confidence to feel that you have conducted your research well and have found out something that it is important.

However, there are some key issues with questionnaire surveys. They are only a 'snapshot in time and space', but this is an important limitation that many students tend to forget. Despite the fact that surveys are well understood by most people, they are not a 'natural' technique, particularly compared with for example interviews or observation, which both involve the type of processes and activities that take place in 'normal life'. As questionnaires are a somewhat contrived situation, responses should be treated with care. Some respondents are unhappy about putting their views 'on paper' (for a postal, or on-line survey) as they may feel that someone they know, or a figure of authority, may read their opinions. Hence, they do not write what they truly believe.

This problem of respondents not necessarily giving truthful answers can be even more of an issue with face-to-face questionnaires. There is a tendency for respondents to give answers that they think the person asking them wants to hear. Commonly, in a face-to-face questionnaire survey situation, the respondent will be unwilling to appear to be in conflict with the researcher asking the questions and agree with them where possible – this is largely a function of human nature. It is termed 'respondent acquiescence' (see Ryan, 1995) and if this happens will clearly have an effect on your results and you will need to build into your survey questions that offset this likely effect. There is also research evidence that women are more willing than men to answer questionnaires face-to-face, but also more likely to agree with the researcher (see Ryan, 1995)! However, there are some respondents who will 'act up' when being questioned. They will deliberately provide provocative or lengthy, detailed answers, because they are acting as if they were taking on a part (in 'a drama', that is not a real-life situation) in which they perform and state things that are not what they truly believe, but nevertheless sound good to them. There are yet other respondents who misunderstand your role as researcher and believe that you are there to help solve their problems. In my experience, such people are frequently elderly – and can be very time consuming! I have, on several occasions, been mistaken for someone from the 'council/electricity company/water authority/gas company' who will fix whatever it is that is no longer working!

Another issue with questionnaires, which has been touched on previously, is that although they can be reliable, they do not necessarily provide useful data for your research project. If you have 100 respondents to your

10-question survey, this may make you feel confident that you have 1000 pieces of data (10 × 100) to use in the 'Results' section of the study and subsequently analyse. However, if each of your questions is simple and straightforward to answer, (which would appear at first glance a plus point) and responses to the closed ended 'Yes/No' type, may look good in your 'Results' section as tables, charts or graphs, there is still the major question: will the questions really provide you with detailed, in-depth, valid answers that can be said to meet your research aims and objectives?

New studies and replicating earlier surveys

Surveys tend to be used for one of two major purposes: replicating previous studies or conducting new studies. It can often be a good idea to replicate someone else's study in your study. There are several advantages to this approach. You will be in a position to use the same overall approach and also the specific techniques. You will be able to draw on the same literature as the original research. You should be able to use the same analytical approach adopted. However, you will need some originality in your own primary research, as this is essential in your research study. Your originality could be that you are a using a similar, but not identical, location for your research. Alternatively, your sample may be made up of respondents with different characteristics. You may conduct your research at a different time of the day, or week or year. It could be that you are updating research that first took place twenty years ago. Whatever the difference, there will still need to be a clear rationale for the choice of your specific research topic. It is not good enough simply to replicate: 'because it looked a good (No, make that easy!) way to complete my study'.

Probably the most important factor, if you do replicate a study, is that you will be able to compare your results, possibly directly with the study that you have replicated. This ability to compare is important in all types of research, but if you have replicated an earlier study, you should be in an excellent position to compare your results. An important point in relation to this comparative element is that your results should be largely similar to the original. Many students assume that their original contribution will be to conduct very similar research to previous authors and yet come up with something very unlike what has been revealed before. This, they believe, is their 'original contribution to knowledge'. Logically, it should be clear that if you have conducted similar research to the study you are replicating, in terms of most of the key research factors, including, for example, approach, techniques, sample, timing and location, but with only one or possibly two differences, your results should be similar. Your 'original contribution' however small, is likely to be because you have slightly modified the study you have replicated. Although relatively small, this can still be your contribution. The results are what your supervisor and examiner will want to see. Even if your results are indeed very similar to the published study you have replicated, you will have added to our understanding, with further confirmation of what has been reported before, and this will be your contribution. If, however, you have found out something very different from the original study, your supervisor and examiner may not necessarily be impressed. They are much more likely to believe that you have made errors in your research, than you are a genius.

'New' studies using questionnaire surveys are usually closely related to previous studies. If they were entirely new, the use of a questionnaire might not be an appropriate approach, as so little is known about the topic that creating a questionnaire would be difficult. Under these circumstances, exploratory research, using inductive techniques would probably be better. However, let us assume that a topic is well-researched, but that the use of new technology such as the internet, or Artificial Intelligence (AI) appears to have changed certain aspects. Under such conditions, it may be possible to modify an older study to build into the original approach reference to the use of the internet or AI. This could still be referred to as a new study.

Conducting a survey

There are a number of ways in which a questionnaire can be put into action. Major ways have been indicated above and include postal, electronic and face–to-face. Traditionally postal questionnaires have been prevalent, but in the past fifteen years or so, electronic surveys, including on-line surveys, have become increasingly common. Even more recently, surveys have been conducted via text messaging, or SMS. Face-to-face is yet another popular way to conduct a questionnaire survey. However, each approach has its advantages and disadvantages.

■ Distributed surveys

Postal and electronic surveys are distributed forms of survey and are still the most common. Both have the advantage of reaching, potentially, a very large number of respondents, which means that you could achieve a good representative sample. However, both suffer from the same disadvantage of potentially low response rates. This can be explained using an example.

If you sent out 10,000 questionnaires and received 1,000 back, you might be happy with this (although this is only a 10% response rate), as you still have a large amount of data in 1,000 responses. But there is a serious issue with this response rate, and it applies to any form of distributed survey. Let us assume (for the sake of argument using easy numbers) that the total population you can access for your research is 100,000, you have sampled 10,000 (this is a 10% sample of the entire population) because that will give you a representative sample but have received 1,000 responses (this is just 1% of the population). Not only is this a very small sample of the population, but what will you be able to indicate in relation to the 9,000 potential respondents who did not respond? Can you say that the 9,000 would have given the same responses in the same proportions as the 1,000? The answer is 'No' and you will not be able to infer anything from the answers of the 1,000 respondents to the potential answers of the 9,000 non-respondents. This is usually referred to as the non-response error (Ryan, 1995). The major implication is that even if you obtain 'good' results from the 1,000 respondents, it will be difficult to generalise your findings from 1% of the whole population, as it is quite possible that it is not a representative sample!

On-line surveys

On-line surveys are now probably the most commonly used technique of survey distribution, and if you are considering using a questionnaire survey then you will probably use one on-line. On-line surveys have been around for over a quarter of a century but have become particular well-used in the last fifteen years or so. There are a large number of strengths of on-line surveys. One particular advantage is their 'global reach' (Evans and Mathur, 2005). As the internet is a global phenomenon, it is technically possible to distribute to a very large number of potential respondents in several different countries/regions. The administrative cost of doing this will be very low in comparison to most other methods of distribution, particularly in relation to postal surveys. In addition, the time taken to send out to a big group of respondents is far less than most other forms of distributed survey. This means that it is possible, in theory, to achieve a very large sample. Online surveys are also flexible in that questions can be modified relatively easily if required and extra surveys sent out to more respondents, without major issues likely to occur. A variety of questions can be used in an online survey and the completion in electronic form means that results do not have to be transcribed and hence statistical analysis can be achieved more easily than if using a printed hard copy of a survey.

However, there are important limitations to on-line surveys. Although it is possible to reach a large sample of potential respondents, there is no guarantee that this will result in a high response rate. In fact, as Evans and Mathur (2005) indicate, response rates can be very low, particularly if the survey is perceived as junk mail and/or ends up in a Spam folder. There are also technological variations (e.g. the Windows PC system and Apple) so that surveys may not appear as they should when they reach the intended respondents. Some of those surveyed may not have the technical ability to accurately complete the survey. This means results can be based on a skewed sample of respondents, most of whom are technically proficient but may not be representative in relation to the survey topic. On-line surveys tend to be impersonal (unlike face-to-face surveys discussed below) and there is also the issue of privacy and a growing concern of whether information provided by respondents will be kept confidential, perhaps used without respondents' knowledge for marketing purposes, or even reach unscrupulous fraudsters who will use it for criminal activities.

Face-to-face surveys

It may be possible, and desirable, to conduct a face-to-face questionnaire survey, particularly if concentrating on a small geographical area. This is very similar to an interview situation, although it may be more structured and take less time. A good location for such an approach is at a festival or an event – in the crowd at a sporting event for example, although it can be done in a main shopping street in an urban area or at a shopping centre. When potential respondents are queuing to see an event, or at a festival, is also a good time to use the face-to-face survey approach. A key advantage of this approach is that a relatively high response rate can be achieved, (usually far higher than distributed surveys), as relatively few people are likely to refuse a direct response to your request to complete your survey then and there, or 'on the spot', as this is referred to. Also, if you are using a particular sampling approach, you will be in a position to adjust it if some potential respondents refuse to answer, and subsequently you should still be able to achieve your desired sample. Clearly, this is not possible with a distributed survey. It will also be possible, in the face-to-face context, to explain questions to respondents if they do not fully understand them. The key disadvantage of this survey approach is that it is likely to be very time consuming, so only relatively few potential respondents can be reached. The face-to-face situation can also present the problems referred to above as 'acquiescent responses' (see Ryan, 1995) with respondents giving the answers that they think the researcher wants to hear. Respondents may wish to complete the survey as quickly as they can to return to what they want to do, so do not necessarily answer it honestly or in full. However, the researcher may be able to pick up on body language signals from the respondents which may, for example, provide confirmation of strength of feeling on a particular issue, or suggest that a respondent is not being entirely honest! This is obviously not possible if using a distributed survey.

Telephone surveys

An intermediate situation between the face-to-face and distributed survey is a telephone survey. Telephone surveys have been traditionally used by marketing organisations for many years. Here, there is obviously direct contact between the researcher and respondents, although the researcher cannot see the body language of the respondent, nevertheless tone of voice may provide further information over and above the precise words used. Telephone surveys can produce high response rates, although it may be easier for the respondent to terminate a discussion before it has finished, than in the face-to-face context. Also, some respondents may not be very comfortable answering questions on the phone, which can affect their answers.

Self-select survey

One other form of questionnaire, which is used by hotels in particular but also some visitor attractions, such as theme parks and zoos, is the self-select survey. In this case respondents decide that they will complete the survey, which is frequently about visitor experience and often requests comments on 'quality' and is therefore a type of customer feedback form. Hotels and other types of accommodation providers often use this approach and will leave self-completion forms in a reception area of the establishment or in the hotel bedroom. Visitor attractions, including zoos and theme parks, may also leave forms for this type of survey close to their respective exit points. Such surveys can be useful, as people who have a strong view on some aspect of the hotel, or attraction, will be able to state this opinion very close to the time they have experience of visiting the location or attraction. A major disadvantage is that response rates tend to be low with this type of research. Also, those who complete such surveys cannot be said to be representative of all visitors. Such surveys usually reveal specific concerns, which is why the respondent has completed the survey. Respondents are very likely to use such a survey when something has not been as it should be. I have completed such a form in a hotel, giving my contact details, when there was an early morning power cut (an unusual occurrence and as I discovered later, not under the control of the hotel), but several weeks later I received a voucher for one night's free accommodation in any one of the hotels in this particular chain. Although relatively unusual, this type of survey is also used by tourism academics, and I have used it reasonably successfully at zoos.

Possible errors arising from surveys

One type of problem associated with the use of a survey 'in the field', the non-response error, has been previously discussed. There are possibly others you will encounter. The most common of these is to do with the individual questions you have used. Whatever you understand by a question, it may be open to interpretation by different respondents. If this is so, then your survey will be far from reliable. To prevent this happening, it is advisable to trial your questions before using them in your survey. The formal way to do this is in a pilot survey in which you test all your questions and modify them if there is misunderstanding, or if there is some evidence that they do not produce useful responses. However, you could informally trial your questions with friends and colleagues. Such trialling, even before the use of a pilot test, is usually known as a 'pre-test' (Mason et al., 2021). This trying out, or testing, is particularly important if you first language is not English. Pilot studies, trials and pre-testing are discussed in more detail towards the end of this chapter.

Another error can arise if your sample is not as you expected or planned it to be. You need to be aware that it is not possible to generalize from your results if your sample is not representative. The nature of sampling and the concept of representativeness are discussed in Chapter 4.

A similar problem to non-response can occur when respondents complete some, but not all the questions, on your questionnaire. If a large proportion of respondents have not answered one question, you will have to make a decision on whether you include the results of that question or leave out the results from this question. If you decide to include the results of this question, then you should indicate what proportion or percentage of the total sample actually answered it, when writing your Results chapter or section. Often it is questions towards the end of your questionnaire that are not answered (this could be due to respondent fatigue) and you should be prepared for this to occur when you design your questionnaire. With my own questionnaire research, I always put the demographic questions at the beginning (although this is not the convention with all academic researchers). My main reason is that I usually want to make use of the demographic information to compare e.g. the responses of different age groups or those of different ethnicity. From my own point of view, there is nothing more annoying than reading an interesting set of responses only to find that the respondent has not provided the demographic information – so I believe there is less chance of missing this out, if it is placed at the very beginning of the questionnaire. Additionally, for a respondent, completing demographic information is a far easier and a less controversial task than having to give your opinion on a contentious issue and this is another reason for putting demographic questions at the beginning. It is therefore part of 'easing the respondent' into the task of filling in your questionnaire. Clearly with a distributed survey, questions can be answered in the order that respondents want to, but in my own experience the great majority of respondents would appear to complete the survey in the order presented to them.

One other related issue is that you may have individual responses that are only partly answered. For example, you may have asked a question requiring a 'Yes/No' answer and followed this up with a supplementary: 'Please explain your answer'. If you wish to discuss responses to the 'Please explain your answer' part of the question, but this has relatively few responses, you will need to make a decision based on how many responses you have received and consider how representative you believe these responses to be.

A problem that occurs in large scale face-to-face research can result from a team of researchers being involved. Each person in the team may have a different way of asking the questions and this can alter the behaviour and consequently the answers of respondents. This will not happen if you are working solely on your own conducting your survey, but frequently students have asked me if they can have someone to assist them in their primary research. This may be because they believe they cannot achieve a large enough sample on their own, or perhaps because they want to have respondents from different locations, so that they are then able to compare in the analysis of the results. My usual response, when asked, would be to suggest that it would be better if the student did all the primary research themselves, thus avoiding the problem.

The covering letter

An important aspect in the use of surveys, and particularly distributed surveys, whether by post or on-line, is the use of a covering letter. This should explain briefly who you are and the nature and importance of your research.

You need to tell your potential respondents that their views on your topic are important. You also need to seek what is termed their 'informed consent' – this means that they are aware of your research and agree to take part in it. You should inform the respondents that you will ensure that their answers remain confidential, that individually they will remain anonymous and that responses will be aggregated in any report to ensure confidentiality and anonymity. You can also offer to send them a copy of the complete results, as an inducement to complete the survey and this may help them have confidence in your indications of confidentiality and ensuring anonymity. Even when conducting a face-to-face survey, a covering letter can be very useful. This is particularly so if it is on university headed paper, as it makes your research seem more formal and more important. When conducting a face-to-face survey, if you have with you a university ID badge as well, this is further confirmation of who you are and can have an (albeit small) effect on the response rate, and it can also help you feel confident in what you are doing. A number of issues in this paragraph are concerned with the ethics of research and this topic is discussed in more detail later in this chapter.

■ Tips for success

Key points to remember when trying to ensure high response rates and achieve useful responses are as follows:

- Use a covering letter that includes the points discussed above.
- Make the questionnaire as easy to complete as possible and also reasonably quick to complete. Anything longer than ten-fifteen minutes is likely to be counterproductive, as respondents may not complete the survey.
- If the survey is distributed, ensure that the respondents know when it has to be returned.
- Try to 'sell' your survey to the respondent (by for example, stressing its importance or, if relevant, referring to its topicality). This can be included in the covering letter.
- It is a good idea to offer an inducement. In face-to face surveys, potential respondents can be offered inexpensive free gifts (such as pens, tea, coffee, biscuits). Those receiving a distributed survey can be offered a full copy of the results.
- In distributed surveys, provide a pre-paid envelope. Respondents are unlikely to want to pay for postage to help you with your dissertation!
- Take some form of ID with you if conducting a face-to-face survey.

The case study approach

Case studies, like questionnaires, are commonly used in tourism and related areas of research. In fact, case studies are used extensively across the social sciences field and in particularly in qualitative research. Students may feel very tempted to use this approach, especially if they are researching one particular example. There are a number of issues with the case study approach, which are discussed later, following a section which considers the nature of case studies and the case study approach.

What is a case study?

Case studies are used to understand what is happening in particular circumstances, at a particular point in time, and attempt to find out why this is happening. As Stake (1995: xi) indicates, a case study is:

> 'the study of the complexity and particularity of a single case, coming to understand its activity within important circumstances'

At first glance, such a definition would appear to fit much research that is conducted in the field of tourism and related studies. Many studies are 'one-offs' and few are long-term or longitudinal. In other words, most studies have been 'snapshots in time and place' (see Mason, 2008). However, it can be argued that if a case study is to be useful in helping our understanding of social phenomena it needs to be more than this. It needs to allow comparison with other studies and literature in the field of study. One of the best-known definitions of case studies is by Yin (1994:13) who defines a case study is an:

> 'investigation of a contemporary phenomenon within its real-life context, especially when the boundaries between phenomenon and context are not clearly evident, and that relies on multiple sources of evidence, with data needing to converge in a triangulation fashion.'

This definition includes a number of key components of a good case study. There is reference to the real-life context – a case study will always be set in the real world and this is part of its value. As the definition indicates, multiple sources of evidence will have to be used and these will need to be compared via a process of *triangulation*. (This process of triangulation is discussed in more detail in Chapter 5.) The definition also suggests both the positive and negative aspects of a case study. The 'real life' means that the case study is grounded in immediate reality. However, this is also the disadvantage of the case study approach, as it means that the results may only be relevant to the specific context and cannot be compared with, and applied directly to, any other situation.

Stake (1995) identifies three significant types of case study. He refers to the first type as the intrinsic study and this involves only one specific study. Such a study does not attempt to make links to other similar cases that have been previously conducted or try to draw wider lessons. If you are thinking of conducting such a study, you should be aware that although you are conducting one case study, it is very likely that you will need to draw upon detailed literature to help design and conduct the study. It also very possible that your study will be able to use similar types of analysis of results, including computer programmes, to similar ones used in the literature. However, be aware of dangers with this case study approach. It is very tempting for you to convince yourself that: 'Nobody has done this type of case study before, so I will create my own research design, research techniques, research questions/topics/issues and analysis techniques'. It is, however, very likely that, although the case study topic seems completely new to you, somebody has conducted similar research – the location, timing, overall context and topics may appear to be very different, but there are, in fact, many similarities. However, because you are convinced that this is a new, original, exciting topic you are equally convinced no-one will have done this before! When it comes to the assessment of your study, you do not want an examiner to say: "You argue your work is original, but a couple of weeks ago, I read a three-year-old article on almost the same topic, although it was conducted in a different place at a different time. If you had looked at this before doing your case study, not only would it have helped you design your questionnaire, but you could have modified specific questions. Also, the article I read would have helped with your analysis". So, in summary, an important message is that you should be wary of believing what you are about to research in a case study is completely new and original!

Stake (1995) indicates that the second type, the instrumental study, unlike the intrinsic study, has as its primary purpose to learn wider lessons. Studying one example, it is hoped, will enable a better understanding of the wider context in which this study is set. It is arguable whether it is better to take a typical study or an unusual, atypical one. In fact, Garrod and Fyall (2013) suggest that an atypical study may provide a greater understanding of a particular topic or issue. They use the example of a holiday where things do not work out and things go wrong, suggesting that this may provide more valuable lessons than one in which things are largely satisfactory.

A third type of study, Stake indicates, is what is termed a collective study. Here, a number of studies linked by a particular context, or theme, are conducted, enabling comparisons and contrasts to be drawn. However, unless you have time and resources at your disposal, it is probably best for you to use either the first or the second type of Stake's case study approach.

Reasons for choosing a case study approach

Most literature argues that the case study approach is best suited to exploratory, inductive and hence qualitative research. It is certainly to be hoped that the results are detailed and in-depth – there is little point in merely 'skimming the surface' when you are only studying one example. As case studies are generally suitable for phenomenological approaches, they tend to make use of techniques, such as interviews or focus groups that will produce detailed, rich data. However, a case study approach can also be used in a deductive, positivist piece of research. For example, this could be where a theory has been developed and applied to large scale activities but has not been tested in relation to a much smaller scale activity (see Brotherton, 2008). This is a good scenario for a research project, in terms of scale and that you can apply the theory at a different level. The actual study would be the testing of the theory in one particular context, different from where it has been applied before. This approach follows the suggestion by Garrod and Fyall (2013) of selecting something atypical to make the focus of the case study.

A major reason for choosing the case study is that what you are studying is inextricably bound up in the context in which it is located. The research I conducted at the Sidmouth International Festival (SIF), which is discussed in more detail elsewhere in this book, had a case study approach (see Mason and Beaumont-Kerridge, 2003). The unusual circumstances in this study are that the Festival took place around the small East Devon coastal resort, at many venues, most of which, (including sports clubs, church halls, cafes, and even the lifeboat station) were used for other purposes for the rest of the year. Therefore, it is not like large scale music festivals, such as Glastonbury, which is concentrated on one site, but spread throughout a town. The general thrust of the research was to reveal the costs of, and revenue generated by the Festival. One aspect of my research was to collect data on visitors' attitudes to the quality of events. It was therefore not appropriate to compare views of SIF attenders with music festival visitors to events such as Glastonbury, where all events take place on one dedicated, and to some extent spatially concentrated, site. At the SIF, the context for the musical event had a major impact on visitors' attitudes to the quality of that event.

However, this discussion of the SIF raises one key issue, of which there are several, with the use of the case study approach. Experienced researchers and students alike may be concerned with how typical, or representative, the case study actually is. Behind this concern is the question: 'How can I know how representative, if I am only studying one example?' A related question is: 'What likelihood is there that something very unusual is happening in my case study?' It is possible that there is something unusual about the nature of the 'sample', or the location where the research is occurring, or perhaps the timing of the research. Yet another issue is what to compare your study to. Should you compare it to another part of your research, in which you have used a different research technique? Or should you compare it to similar studies in literature? The answers to these questions, which are related, as they are largely concerned with 'representativeness' of your case study, are threefold. First you are studying just one example to gain deep insights; second that the context for the research of the study is vital to the understanding of what is happening; and third that it is not really your intention to generalise from this study to others. However, having said this, it should still be possible to compare your study to similar types of works in the literature.

A solution to a number of these issues could be to conduct several related studies in different places and times. You could also vary the techniques previously used. However, this solution will inevitably be time consuming and this to some extent negates the approach of using one case study which you want to research in detail and in-depth. Nevertheless, there are a number of variations on the case study approach (see Yin, 1994) in which multiple cases can be investigated and compared. However, given the time frame and resources available for your dissertation or thesis, this type of case study approach is likely to be more problematic.

Validity and reliability

These two terms are important concepts in research in tourism and related areas. Unfortunately, outside of the academic world, they are used in 'common speech' but rather loosely and often as if synonymous. However, in research they have very specific meanings.

■ Validity

In relation to the concept of validity, I like to think of this term as being almost synonymous with the term 'appropriate'. One important question that can be asked of any research project is: Does this project really address the issues it claims to? This is a question about validity. Another way of putting this question is: Were appropriate methods/techniques/questions used to achieve the aims and objectives? For example, take the situation in which a researcher has, as a research aim, to find out about the motivation of visitors to a particular theme park. If we assume the researcher asks the question of visitors via a face-to-face questionnaire, used in the car park as they are leaving: 'Have you enjoyed your visit to the theme park?' In addition, the researcher asks: 'Will you be returning again to visit the park?' These two questions may appear at first sight to be clear and logical. It seems likely that visitors will be able to answer them without much difficulty, so it would appear that they must be appropriate. But are they *valid* questions given the *aim* of the research? The answer is 'No', although you may be thinking: Why not? The reason that they are not valid relates to the research aim. As indicated above, this is to find out what motivates visitors to go to the theme park. Asking visitors if they have enjoyed the visit and will they be returning, may generate data, but these answers do not address the aim, as they are not likely to reveal the motivations for visiting the theme park. The two questions asked are both likely to have 'Yes or No' as their possible response categories. The aim requires finding motivations (reasons for visiting) the theme park. 'Yes or No' responses will not do this with these questions. So, the questions are not valid in relation to the aim of the research. However, if the researcher had presented a series of statements such as: 'I visited many years ago with my parents and wanted to bring my children here', 'I like to visit with friends and family', 'It is good value for money', 'I like the white-knuckle rides here', and 'Children are well catered for', 'and provided 'Yes/No' response categories, then these would be valid questions, as the researcher is providing a list of possible motivations to which visitors can respond.

Internal and external validity

There are important sub-categories of the concept of validity in research. When you write the final part of your dissertation, particularly if it is a quantitative study, you are likely to be suggesting something to the effect: 'My results indicated that... and, comparing my primary research with important literature... the main findings are similar to the study carried out by... although they are not quite the same as the research of..., but based on all of this, the main conclusions are...'. However, if you do not have evidence from your study to support what you have written, then your examiner is likely to comment that your dissertation lacks *internal validity*. Internal validity relates to how the arguments used in the dissertation or thesis justify the conclusions (Long, 2007). This means that there should be a logical link between your primary research and what you write in your conclusions. If you then go on to write that you have revealed important factors in your study, which will almost certainly apply to other similar circumstances, this relates to what is termed, *external validity*. External validity means that you can generalise your primary research results to other situations where conditions are similar to the conditions under which you conducted your study. However, if your research, as in the example above, lacks internal validity, it will almost certainly not have external validity.

Construct validity

Other important terms are linked to the concept of validity. One important term is construct validity. To a large extent this is to do with whether you have selected the overall, appropriate way of doing what you intend to do (in particular, using the appropriate methodological approach) in your research, according to your aims and objectives. If you want to find the reasons, in the actual words of respondents, as to why they prefer visiting the World Heritage Site (WHS) of Stonehenge, rather than the WHS at nearby Avebury, then using closed-ended questions on a face-to-face questionnaire, will not be an appropriate way to do this. Using interviews, or possibly open-ended questions on a questionnaire, would be appropriate ways. In the above example, it is relatively easy to see where an invalid approach has been used. Construct validity is likely to involve very careful consideration of whether your selected approach really will measure what you intend to measure. In your study, trying to make sure you have construct validity, may involve comparing your approach with what other researchers, who have conducted similar research, have done.

Content validity

Content validity is concerned with the actual content of the questions you have posed. It is to do with whether specific questions address the key issues that are stated in the aims and objectives. You may have construct validity – the appropriate technique – but some of your questions may not be in a logical order or provide adequate coverage of your topics. So, in this case you would not have content validity. In the earlier example in this section, the study of motivations for visitors to a theme park, the discussion is, strictly speaking, concerned with content validity. The construct – using a questionnaire survey is appropriate (so you have construct validity) – but the actual question contents, as above, are not appropriate (so you do not have content validity).

Reliability

Reliability, at first glance, may be for you a term that means almost the same thing as validity, but in research it does not. A good synonym for reliability is *consistency*. Reliability refers to whether a measure is consistent over time and when used in different contexts (see Brotherton, 2008). It can be said that reliability is about replication – being able to repeat the same results (Clarke et al., 1998). Reliability is very important in your study, because if you have doubts about a particular measure, you will not be able to make strong claims about what you have revealed in the primary research. One of the key areas in which you may have concerns about reliability could be in relation to attitude scales. Is it better, for example, if using a Likert scale, to have five points or seven points on the scale? A similar question could be posed in relation to the bi-polar semantic differential scale. The question in each case is concerned with the consistency of the scale in measuring respondents' attitudes. What can be said in answer to the question, is that if the scale is used repeatedly and produces consistent results, then it can be considered reliable.

Ensuring reliability

How can you ensure that the specific questions you have used are reliable? If you are largely replicating a previous study and have used an almost identical question to the one in the study you are replicating, and have achieved very similar results, then you can claim that your results have a high degree of reliability. If this is not case, then you have a very good rationale for the use of a pilot study. Details of how to conduct a pilot study are provided in the next section of the chapter. A pilot study is usually conducted prior to the use of the main research instrument, but you can conduct what is known as 'test and re-test' (Clarke et al., 1998) in your main research project. This is most frequently used within experimental research, to refine specific questions. However, it may be difficult within your research, particularly if you have sent out a postal questionnaire. If you are using an electronic online survey, this could be modified on the basis of the test/retest approach. Researchers may also conduct a pre-test, before using a pilot study (see Mason et al., 2010) in an attempt to achieve as great a reliability as possible.

There are a number of ways of statistically measuring reliability. One relatively easy way is known as the 'spilt-half method' (see Clarke et al.,

1998: Ryan, 1995). The sample you have used is divided into two equal sized parts and then compared. The usual way of doing this is to put all even numbered statements in one group and all odd numbered statements in the other. If the survey is reliable, then the sample's scores on each of these two artificially created groups should correlate. Details of how to conduct this test are provided in statistical texts, and also for example in Clarke et al (1998) and Ryan (1995).

Research Design

	Appropriate	**Inappropriate**
Data Collection	Research has validity and reliability	Research lacks validity, requires redesign and implementation
	Research lacks reliability and needs to be implemented again	Research lacks reliability and validity
	Inappropriate	**Inappropriate**

Figure 3. 1: The relationship between validity and reliability in research design and data collection. (Source Ryan: 1995)

Figure 3.1 gives an indication of the importance of validity and reliability. Notice that of the four cells, only one, the top left-hand corner indicates that both research design and data collection are appropriate. The cell immediately to the right of this one indicates a problem with the research design, (i.e. an issue with validity) therefore the need for a redesign before implementation. The lower left-hand cell has problems of reliability with data collection which means it needs to be implemented again, whilst the lower right cell has problems with both design and data collection so indicating issues with both validity and reliability.

Finally in this section, bear in mind that although validity and reliability are linked and vitally important concepts in all research, but particularly quantitative research, there are some important aspects to remember about the relationship between them and their differences. You can have a reliable measure, but it may not be valid. My question above, 'Have you enjoyed your visit to the theme park?' in the example of 'motivation to visit', could produce reliable results (i.e. respondents give consistently truthful answers to the actual question), but the question is not valid, because it does not reveal information about the respondents' motivations, which is the aim of the research. You can also have a valid measure, but it produces unreliable results. For example, you could ask, via interviews, fifty restaurant diners about 'service quality', but the results may not be reliable because of the different understandings of the concept of 'service quality' amongst the diners. So here, the method is valid, but the results are not reliable.

Conducting a pilot study

Before you undertake your main research project, particularly if you are about to involve a number of participants, such as visitors to an event or an attraction, or to interview some key players such as managers of a sport stadium, or a festival organising team, it is very important to consider what could go wrong – this is in order to try to ensure that you get it right. Therefore, you need to make sure that you do the best that you can, but, as it is not possible to guarantee that your research will be successful, you should have a contingency plan. What you should be aware of also, particularly if you are doing a PhD, is that your examiner will be interested to know how you coped when things did not turn out as expected. This happens to many PhD students and, what they do following this problem is all part of the research process and, as indicated in Chapter 1, should be seen very much in the context of 'researcher as apprentice.'

Even if you are not doing a PhD, it is possible and, in some cases highly likely, that you will have to make changes to your original conceptualisation of the conduct of your primary research. and you only have a few months to complete your research. But how can you ensure that your research process is the best possible at the time that you are doing your dissertation? Clearly it is not possible to give a definitive answer to this question. However, an important way to reduce the possibility of things going wrong is to try out your approach before you really get started on the main research process. In some types of research project, it is possible and desirable to conduct what is known as a pilot study. A pilot study enables you to try out a number of aspects of your proposed research. These include, for example, actual questions you intend to use, the type of respondents you wish to involve, the type of location where you want to carry out your research and the time of day you plan to use. It also may be used to make decisions on the nature of your sample, in terms of for example, demographic factors, or whether it would be better to use one form of sampling rather than another.

However, conducting a pilot study takes time and the results from the pilot trial have to be analysed. Given that your project, particularly if it is a dissertation, usually has to be completed in a relatively short time frame, this often puts students off from conducting a pilot study. Nevertheless, this should not be a deterrent, as a pilot study frequently helps to save time later in your research. In extreme cases, it may also prevent you from conducting research that will not be good enough to enable you to produce a successful project.

The most important factor to be aware of when conducting a pilot study is not what you find out about your proposed topic, as the pilot study is not designed to provide answers to questions that can be used in the analysis section of your study. It is intended to enable you to learn about your proposed research approach, overall methodology and, in particular, specific questions and techniques that you intend to use. In extreme cases, you may find that what you intend to use as the overall methodology and hence the techniques you have proposed, will not work. If this occurs, it is far better that it does, before you really get started, than when you are actually carrying out the main part of your research project. You do not want to hear your supervisor say: 'If only you had conducted a pilot study, you would have known that ... would not work.'

However, if you have carefully considered the key concepts of ontology and epistemology and linked this to your proposed methodology and individual research techniques, it should be the case that, following your pilot study, you are only required to make some relatively minor modification to, for example, some questions on your questionnaire.

■ Pre-testing

Before even conducting a pilot study, it is well worth considering a pre-test of your research approach. In particular, this can be very useful to test out particular questions you wish to ask but could also be used to test a particular research technique. For example, you may need to decide whether a questionnaire with a large number of closed-ended questions, would be better than one with a large number of open-ended questions. You could design an example of each, containing the type of questions you regard as appropriate for your topic. It is not essential that you conduct a pre-test with an actual sample of the type of respondents you intend to use in your main research. Instead, pre-testing could involve you giving a copy of each questionnaire to, for example, friends or fellow students, or just people you consider are able to provide useful comments. You would then ask them to consider each survey and give you feedback in the form of evaluative comments. This should help you decide, before you then follow this up with a pilot version of the particular survey you have decided to use.

■ The pilot sample

When you use your pilot study, you should try to conduct it with the same type of respondents as you will use in your main study. For example, if your sample will only include those aged between 21 and 40, then it is not a good idea to test it on exclusively those who are all over 60 years old. You should only need a small number in you pilot study (e.g. if your proposed sample total is 100, then a pilot of 10 – 15 respondents should be sufficient). It is possible that you have to modify several questions as a result of the pilot. If this is the case, then you may need to test the revised questions again and this may involve another pilot. However, if your questions are well designed to begin with, then it should be the case that only minimal changes are needed. A very important point to consider when conducting the pilot survey, is whether or not you should tell your small group of respondents that they are doing a pilot study. It is possible that they would not understand what you mean, if you did tell them, but if you were to explain what you are doing, then possibly they may be reluctant to take part. Why should they bother to take part when you are not interested in their responses, but only whether your questions work? This could be their reasoning. However, some people are willing to accept that they are doing a pilot study to help you try out your research technique. Whatever you decide to do you will need to justify this in your dissertation or thesis.

You may be now thinking that pre-testing, let alone conducting a pilot study, will require a lot of time and effort, and is it really worthwhile? The simple answer is that time spent at the *beginning of the study* will *save you time later on*. If you have good, clear, answerable questions, then you should have good, relevant data to present and analyse. This may not happen if you do not conduct some form of 'trial' in advance of doing your main study.

Ethics

Ethics is concerned with moral values, and principles such as integrity, honesty and transparency. In particular, it is concerned with how we relate to other people and with ways in which we behave. Your initial reaction to this may be to think: 'What has this to do with research?' In response to this question, it relates largely to the way in which you interact with the respondents in your research. It is also concerned with how moral principles inform the design, conduct and publication of research (Brotherton, 2008)

It is very likely that you will have some form of contact with specific sets of people in your research. If the respondents are part of an on-line survey, even though you do not have direct contact with them, you will need to consider ethical factors. This is because it is possible that you could inadvertently reveal some confidential information about one or more participants. If you are conducting face-to-face interviews, a focus group or are using an 'on-the-spot' questionnaire survey, there will be a very important ethical dimension in each of these circumstances. If conducting observational work, where you may not make direct contact with the participants in the research process, the ethical factors are still important. Even if the main source of data is documents/media sources – in other words your sources are largely secondary – there is still a strong likelihood that you will need to consider the ethical dimension.

Hence, all students working in tourism and related fields, whatever the topic or chosen methodology, will need to consider ethical factors in their research. Most universities require students to complete an ethical statement, and this will usually be on a form that the university produces, which may be largely a checklist, but possibly requires students to write in their own words details of their study that have ethical dimensions.

Why ethics matters

Why is a consideration of ethical factors in social science research so important? You may already be aware that ethics is particularly relevant in research when, for example, the project involves medical factors that clearly have a dimension in which respondents could suffer through being identified. This could be the case if respondents had a serious medical condition and the researcher passed on this information, (which should be kept confidential) to, for example, a drugs company or the media. You may also have heard that certain vulnerable groups, including children or disabled people, need to be the subject of careful consideration in relation to ethical statements.

This may have led you to think that there is nothing in your dissertation research that is in any way connected to ethics. This is unlikely to be the case, however. All research involving people as subjects of the research will require ethics to be considered and some form of statement in relation to the ethical factors prepared.

Assuming you are not researching with vulnerable groups and also not putting yourself in a dangerous position when researching, the most likely issues you will need to consider are confidentiality and anonymity. You will need to seek and obtain *informed consent* from your respondents, for example, if you are using a questionnaire survey or interviewing respondents. What this means is that you will need to inform them in advance, of the nature of your research and let them know what will happen to the information that you gather, including what they tell you. If you are doing this in a face-to-face situation, then you can read out a statement to the respondents. This could be written in the form of a covering letter that accompanies your research. This covering letter can also contain information about you, your university, the fact that you are doing research as part of your degree, the aims of the research, as well as what will happen to the data you collect. In postal or on-line/email questionnaire surveys, the covering letter can accompany your survey.

Protecting confidentiality

An important issue that you will need to address is where and how the information you gain will be stored – to ensure it is secure. You will also need to inform participants that their personal information, if they have provided this, will be kept confidential and in terms of presenting results and findings will be aggregated, so that it is not possible to identify any one individually. Even though you may not have taken the respondents' personal details (their name and address for example) it may still be possible to identify them. For example, there may be something about their age, gender, education level, nationality or ethnic origin, or any combination of these, that would enable anyone who saw your data to identify them from this information alone.

If you are conducting interviews or a focus group, you will also need to inform participants that when you write up your research, it will not be possible to identify them individually. The important concept here is anonymity. Anonymity means that all your data is in effect anonymous data – it is not possible for anybody to identify any one individual involved, as a respondent in the research. The only person who should be in a position to identify an individual is you, as researcher. You need to maintain a formal, professional role when conducting your research and this is very important in terms of the ethical position. Do not tell your friends something to this effect: 'I just interviewed this woman in the High Street, she was really weird, seems to hate all foreign tourists, wants them to be sent home, or not even allowed in the country. She told me her son got in a fight, about football I think it was, with a Spanish tourist down at the pub on the corner and broke his jaw and the son is now facing a big fine and maybe a prison sentence! She lives somewhere in Acacia Avenue she told me'. There is sufficient information in 'what you told your friends' to fairly easily identify the woman who was interviewed!

When it comes to presenting your results, to ensure the continuation of confidentiality and anonymity, you will need to come up with a system that you follow in your presentation and also the subsequent discussion and analysis. A common approach is to use pseudonyms. This means that you invent names for your respondents. So, Mr. George Smith becomes, for example, Colin. Using the name Colin, it will not be possible for George Smith to identify himself or more importantly anybody else to do this. Some researchers use a system in which interviewees are numbered, so when presenting the results would write, for example: 'INT 1 stated that….., while INT 5 indicated……, however INT 2 did not agree with INT 1, but INT 3 and INT 4 did and their views could be summarised as…'

It is also important to be aware that respondents in questionnaire surveys should be provided with anonymity and their responses kept confidential. It is usually the case that results from a questionnaire survey are aggregated and it is therefore difficult to identify individuals, but in some cases, particular individuals' responses could be of great interest to you as researcher, because for example, they are very different from all others, or suggest a topic/theme that you believe is very important and you intend to discuss in your analysis section/chapter. Under these circumstances it is possible that you accidentally reveal enough information for them to be individually identified – you should ensure that this does not happen.

An important ethical issue, which may not have occurred to you, is your own safety. You need to consider the question 'Will I be at risk when conducting my research?' The answer is most likely to be 'No'. However, you need to give this a good deal of thought. For example, if you plan to investigate, through observation, the use by tourists of space on a beach, in terms of the activities of individuals, couples, groups and families, you may initially assume this carries no risk to you and will be quite straightforward. However, if you have decided that you do not wish to be identified as a researcher, (this is covert observation and is discussed in more detail in Chapter 5) but will photograph and/or make a video recording of the activities, to help with your research, you may find you are in conflict with those you observe and in certain circumstances may be breaking the law. In the UK it is illegal to photograph or video those under 16 years old in a sporting context, for example, without their parent's permission. One reason for the relatively recent introduction of the law relates to alleged paedophiles photographing/making videos of children. Clearly, you do not want to find that you are putting yourself in a position where you could be accused of being a paedophile!

At some UK universities (and also at those in other countries), students are required to complete an ethics form prior to the start of their primary or field research and this will cover issues discussed above. Hence, it is possible that you will be required to complete such a form in your own institution.

Writing this section in your study

Research philosophy and design will form a very important section of what will probably be part of the chapter in your study concerned with, and usually title, 'Methodology' and commonly follows the 'Literature Review' chapter. This is nearly always the case if your study uses quantitative approaches. If however, you are using a qualitative approach the chapter on your methodology may have a different location in your dissertation, but it will have to be written wherever it is located.

A key element of the 'Methodology' chapter will be to explain your choice of research philosophy and its related paradigm. It will not be sufficient to merely describe the philosophy and related paradigm and also the overall research approach, and it is of particular importance that you provide a detailed rationale for your decisions here. When you write this rationale, you should make use of the extensive literature on quantitative and qualitative research philosophy. Students frequently provide a summary of key terms including ontology, epistemology and methodology, what they each mean and how they differ in the two major research paradigms. However, in this discussion you should ensure that you concentrate on the research approach that you have used and not write at length about the 'other' research approach – it will not be strictly relevant to what you have done.

The rationale for your choice of research approach should be a critical examination of the particular research philosophy and overall approach that you have used. This will involve a discussion of the advantages and disadvantages of the particular approach you have employed. This discussion should stress, in particular, the advantages of your chosen philosophy in comparison to what you have not selected.

If you are using a quantitative approach and are largely replicating another researcher's approach and techniques, then you will not need to discuss in great detail, the research philosophy you will use and the concepts of ontology and epistemology, as these should be clear from the work of the researcher(s) whose work you are replicating, although in a different context. However, if you are using an overall research approach and techniques that are different from what has been used before in relation to your topic, then you will need to carefully indicate why you are doing this. Your rationale for the approach you have adopted, in this case, is likely to need discussion of the important concepts of ontology and epistemology, as well as methodology.

Having discussed the rationale for your choice of a research approach, you will also need to discuss the nature of this approach, be it a case study or a questionnaire survey, in particular, in terms of design and conduct of the research. This discussion will also require you to make use of the large range of literature in the social sciences generally, but also the more specific material concerning tourism and related fields.

In addition to your rationale and the nature of the particular approach you have used, you should include reference to the implications of your selected research philosophy for the different techniques you could use. If you have used a quantitative approach, there is discussion at the end of Chapter 4 on how to present material in the Methodology chapter of your study on the different research techniques you have used. However, if you have used a qualitative approach, there is discussion at the end of Chapter 5 on how to present material in the Methodology chapter on the different research techniques you have used. If you have used a mixed methods approach, there is discussion at the end of Chapter 6 on the different techniques you may have used.

At some point in the 'Methodology' chapter (or possibly in another chapter) you will need to discuss ethical issues. If required to complete an ethics form by your university/institution, it is very likely that most of the information concerning ethical issues and your own research can be included there. Nevertheless, it is important to make some reference to ethical matters, if only to indicate that you are aware, at least of issues to do with confidentiality and anonymity. If there are other ethical issues, such as to do with the risk to participants or indeed your own safety, these will require more detailed discussion.

Student activities

1. What do you understand by the following terms: ontology, epistemology and methodology and how are they related?
2. In which research paradigm will your study be located?
3. Under what conditions would you use a hypothesis?

4 With reference to Case Study 3.1, which approach would you use: quantitative or qualitative? Explain your decision.
5 Why are theories important in tourism and related research?
6 What is a dependent variable and what is an independent variable?
7 What do you understand by the terms validity and reliability?
8 Why is it frequently important to conduct a pilot study?
9 Why is important for all students in the tourism and related fields to have to consider ethical dimensions in their research project?
10 In relation to ethical factors, under what circumstances would you need to gain permission to access the site(s) of your research? Why is it important to gain permission?
11 How would you store information gained in a face-to-face questionnaire survey? What would you tell participants about the storage of this information?
12 Key ethical terms relating to the use of any human subjects (people) in tourism and related research are confidentiality and anonymity. Explain what you understand by these terms.
13 Why is it necessary to use pseudonyms or a coding system if using interviews or focus groups?
14 Under what circumstances could you be at risk as a researcher? In relation to the response to this question, what could you do to reduce the potential risk to you?
15 Why is it necessary to consider ethical factors when conducting observation, which you intend to do covertly, and which therefore will not involve direct contact with participants?
16 Consider the following scenario. You live in a small coastal town that attracts large numbers of tourists during the summer season. Newspapers report that petty crime increases during the tourist season and the media generally suggests that police make more arrests for shoplifting and pickpocketing, at this time of the year. Consider how you would conduct research with the provisional title: 'Tourism leads to an increase in petty crime during the summer season' and with the objectives to find out the following: if petty crime actually increases, if so, what type of crime this is, who the perpetrators are, who the victims are and what the consequences of the crimes are. What important ethical aspects would there be if you were to conduct such research?

4 Quantitative Research Techniques and Conduct

Introduction

As has been discussed in previous chapters, one of the key research paradigms is that of positivism. This earlier discussion has also indicated that those involved in positivist research almost always make use of a deductive research process and their research is most frequently quantitative in nature. In Chapter 3, the overall approach of the use of surveys has been discussed and as this chapter indicated, the most common type of survey used in academic research in tourism and related studies is the questionnaire survey. The great majority of students involved in preparing and writing a dissertation or thesis will be using a quantitative approach. In fact, as noted earlier, a good number of students, when asked what research is, will answer that it is using a questionnaire survey! This chapter focus on the use of various techniques for collecting quantitative data.

Questionnaires

■ Designing questions

Although by definition, questionnaires make use of questions, there is a large range of question types – unfortunately not something always recognised by students. How you generate the questions is important, more important than many students recognise. It is hopefully relatively easy for you to come up with lots of questions to do with your topic, but you need to know that they will actually provide useful data. This, by the way, does not mean your questions will give you the answers that you expect you should receive, based on your own views, but will yield answers that make sense, are reliable and can be analysed. Can your questions actually be answered, are they ambiguous or are they leading questions? These are some of the points you need to consider, and you need to think about these before actually putting your questionnaire together.

One way to assist in the questions-creating process is to have your aims and objectives in front of you at all times when you are doing this. You may have an elegantly worded question that you are keen to ask respondents, but you should think: 'Is it relevant in relation to my aims and objectives?' If the answer is 'No', then what is the point of asking it? A defence of 'I thought it was a well written, interesting question' is not acceptable. It will waste the time of your respondents and also may mean that they do not answer other questions that really are linked to you aims and objectives.

If your study involves applying a theory and testing hypotheses, it may be the case that you are making use of questions previously used by another researcher. This is an acceptable way to create your questions, providing you acknowledge their source. Many of the questions may work as well as in the original, but some others may need to be modified for the context of your research. However, if you are conducting new research, or large parts of your research topic require new questions, then you will have to give considerable time and effort to creating these questions.

One very important way to generate questions is to use the literature on your topic. Your initial reaction to this, could be to think: 'There is not much literature on my topic'. However, as an example, if we assume you are investigating attitudes of visitors to a heritage attraction that, as far as you are aware from your literature survey, has not been studied in an academic way before, you should still be able to find a heritage attraction, somewhere else that is similar and where academic research has been conducted. If research has been conducted, then you could modify the questions used in the study of the similar attraction for your research context. This is perfectly acceptable, but only as long as you acknowledge, giving a reference, where you have found the original questions. This also has the advantage that you can compare the responses you gain in your primary research with the original study you used to create your questions. If, however, you find literature that is relevant, then use all or just some of the questions, but do not acknowledge the source of the questions (the name of the author (s), and the date of publication), then you are committing plagiarism and this could lead to serious penalties, including being asked to produce a complete new research project, or worse being thrown off your course.

Your questions can also originate in material that is not academic. For example, the managers of a tourism facility may have conducted their own research. You can 'borrow' their questions, once again providing you acknowledge this. You will then be in a position, similar to that outlined above, to compare your results with those you have 'borrowed'. Yet another source is topical stories/reports in newspapers, on TV, or on-line. For example, I conducted research at Stonehenge in 2004 and some of my questions were based on a report by UK MPs into the 'quality of the visitor experience' there. Remember that it is essential that you should acknowledge where your questions came from – you are expected to do this anyway when writing up your study, but also it indicates that you have read around your topic and whatever you use should add some authority to your overall questionnaire design.

There are a number of software packages that are available to help you design and create your questionnaire – your institution is likely to have a range of these, but if not, or you wish for something your institution does not have, you will probably have to purchase a copy. There are also on-line questionnaire design sites, and these may allow you to have a trial run with your question design. One such site, that has been in existence for several years and been used by several students I have supervised, is SurveyMonkey, and as indicated in Chapter 3, there are certain advantages to using on-line surveys in questionnaire-based research. In fact, there are now a very large number of organisations, in addition to SurveyMonkey, that provide information on how to create online surveys. Many of them use a 'survey builder' for you to create and edit surveys quickly and directly from an internet browser. It is possible to use templates created by the organisation or to create your own template. This will allow you not only to design questions but also arrange them in a sequence that suits you. Some on-line packages can also calculate response rates to your survey, reveal trends in responses, create reports on your results, analyse your results and even send out reminder messages to your intended respondents, who have not answered by your set deadline. The product of using such online organisations can be an attractive, professional looking survey, which may lead to a higher response rate. However, you may find that you have to pay to make full use of the online survey tool.

One of the key aspects when constructing a questionnaire is to be aware that although it is made up of several different questions, they should be linked to its main theme. This means when it comes to analysing the questionnaire, it will be possible to gain an overall perspective on what each individual thinks. This will be achieved by comparing all the responses of each one and should enable you to write 'This individual tends to have the following views...' with a summary of their responses. However, given there are several questions on the questionnaire, it will be possible to compare all the answers to one specific question from everyone involved in the survey. This is frequently the way questionnaire research is discussed, taking in turn each question and a summary of its responses, possibly via tables, graphs or a statistical presentation. Combining these two approaches, (all answers given by one respondent and all respondents' answers to one specific question), also means it will be possible to write, something to this effect: 'Despite this individual generally have the following views in relation to Questions 5 and 7….., they agree with the majority of respondents on other questions'.

Closed-ended and open-ended questions

The type of questions that are associated particularly with questionnaires are closed-ended questions. These restrict responses to a limited range of pre-coded categories. A very common type is the simple choice of 'Yes' or 'No'. Such questions are very easy to analyse, and the results can easily be presented as, for example, percentages. However, there are other types of closed-ended question which are more complex than this. These will have a number of categories from which respondents need to select. Such questions will usually have the possible responses enclosed within a table, figure or box and there should be a space near each possible response category for respondents to put a tick or indicate by some means what their response is. A key factor when designing closed-ended categories is to ensure that each is mutually exclusive. This is easy enough when there are only two possible categories such as 'Yes' or 'No' but can be more difficult with several. Below is an example of how **not** to write a question intended to have mutually exclusive categories (see Table 4.1), followed by how the same question information **should** be presented (see Table 4.2). If you want to find out the age of your sample, **do not** ask respondents to fill in a table such as Table 4.1.

Table 4.1: How not to use mutually exclusive caegories

Please use the table to indicate your birth year:

Before 1940	
Between 1940-50	
Between 1950-60	
Between 1960-70	
Between 1970-80	
Between 1980-90	
Between 1990-2000	

Why is Table 4.1 not easy for respondents to complete and therefore analyse? Well consider if you were born in, for example, 1950, or 1980, how should you complete the table? You would need to put a tick in two boxes! So, in these cases, the categories are not mutually exclusive, which is what they should be. Table 4.2 below indicates how you should present this question.

Table 4.2: How to use mutually exclusive categories

Please use the table to indicate your birth year:

Before 1940	
Between 1941-50	
Between 1951-60	
Between 1961-70	
Between 1971-80	
Between 1981-90	
Between 1991-2000	

Whatever questions of this type you are asking, you must have mutually exclusive categories. Such questions should also have categories that account for all possible answers – they are known as mutually exhaustive categories (Long, 2007). For example, let us assume you have a question about how people travelled to an event. Depending on your aims and objectives, you may be interested in particular types of transport, but there is always the chance that you have not included every possible means of transport, so you table should look like Table 4.3. The important category here in terms of this discussion is 'Other'. Asking such a question means that all possible means of transport have been included. However, it will not be a very good question if the category with the largest number of responses is 'Other'! So, you need to be aware that although you should have an 'Other' category, in whatever topic you are asking questions, you should try to ensure that you cover all the possibilities with separate categories, so that 'Other' records very few responses.

Table 4.3: An example of mutually exhaustive categories

By Car	
By Bus	
By Train	
By Motor bicycle	
Other	

Open-ended questions are, in effect, the opposite of closed-ended questions, in that there are no pre-set categories. Respondents are being asked to give answers entirely in their own words. However, there is a need to indicate how much information you wish to receive. A good way to do this is to provide a set space on the questionnaire. It is also helpful for respondents if you give them ruled lines under/adjacent to the actual question. If you provide three lines then this is an indication to the respondent how much information you wish that they should provide, only one line and this suggest to the respondent that you would like a shorter answer.

Why use open-ended questions on a questionnaire? Such questions are to be used if you do not really know what categories you can put – in other words you do not know the range of possible answers. This is particularly the case, when there may be a topic, or parts of a topic, that you are not sure how respondents will answer. Therefore, you wish for respondents to express their views entirely in their own words.

Another rationale for the use of open-ended question is to confirm (or not) that a closed-ended question is reliable and therefore your overall questionnaire is reliable. I have asked both open-ended and closed-ended questions on the same questionnaire and each of these types of questions was on the same theme. My open-ended questions were placed at the early part of the questionnaire and the closed-ended questions followed on later. The rationale for placing the questions in this order, with open-ended ones near the beginning of the questionnaire was so that respondents were less likely to be influenced by the pre-set response categories in the closed-ended questions, when giving their responses to the open-ended questions. The questions were concerned with attitudes to a tourism development and the closed-ended questions were used, partly, to check if the open-ended questions were reliable. In other words, using one example, if the open-ended responses indicated a positive attitude towards the development and the closed-ended questions revealed very negative views, then there appeared to be a problem. In fact, in this research, a very large number of the closed-ended responses were very similar to the open-ended answers, so this made me believe that my questions were generally reliable, and I claimed this in the article I submitted, based on this research, that was later published in a tourism journal.

However, a key issue with open-ended questions is that they will be potentially difficult to analyse. Allowing respondents to say whatever they want in response to your question means potentially an almost infinite range of possible answers. Nevertheless, in reality, you are more likely to find that, despite your open-ended question, responses are restricted to a relatively small range of possible answers. This means that your analysis may well be much simpler than you at first believed.

The open-ended question approach also can be used deliberately as a kind of 'balancing act' set against the closed-ended questions approach. The rationale for this would be to make the respondents feel that they are more in charge of the research process and have the freedom to give their own responses, at least in relation to some questions. In terms of the analysis of open-ended question responses, you will usually have to create categories on the basis of the material in the responses. The advantage of this is that, as with the responses of closed-ended question, coded responses to open-ended questions can be quantified and therefore analysed as such. This process of converting open-ended question responses into analysable quantitative data is discussed in Chapter 9.

There is also a compromise position in which you ask an open-ended question, but you have decided on the response categories in advance. This may seem to run counter to the principal of open-ended questions leading to open-ended responses. Nevertheless, depending on the actual question, it is quite likely that there is only a limited range of answers possible from your respondents. However, you would probably need to pre-test your questions and possibly conduct a pilot survey to establish the categories. You may be wondering why it is worth using this approach. One rationale is that despite your pre-testing and piloting, you are still not confident that you know all the possible responses. Nevertheless, you are fairly confident that almost all of the respondents' answers will fit into the categories you have pre-determined. However, the decision to present the question as open-ended, is because you are not one hundred per cent certain of possible responses and want to ensure that you do not omit potentially important answers. In a similar way to closed-ended questions, as the categories have been pre-determined, then quantitative analysis will be possible.

There is another important type of closed-ended question that are commonly used on questionnaires and frequently used by students in their research projects. These are attitude-scale questions and include, the particularly well known, Likert scale, but also semantic differential scales. Before discussing each of these, it is important to be aware that there are different types of measuring scale used in social science. It is usually accepted that there are four types of scale: nominal, ordinal, interval and ratio (Brotherton, 2008: Long, 2007; Veal, 2011). The following is a brief discussion of each type of scale to explain what it is and how it differs from the others.

Measurement scales

■ Nominal scales

Nominal scales are often used on a questionnaire to name or classify respondents' characteristics. Hence, classifying respondents as male or female is employing a nominal scale. Likewise, putting respondents into different age groups is also employing a nominal scale. The information in such nominal scale is relatively simple but can be very useful in terms of details about the sample. The nominal scale information is also important, particularly when comparing responses to specific questions on the questionnaire with nominal scale information. Hence, when conducting the analysis it may be possible to state for example: 'In relation to Question 7, male respondents indicated ......, whilst female respondents.......', or 'In relation to Question 8, those in the age category 21-30 generally had very different views to those in the age category 41-50.'

■ Ordinal scales

Ordinal scales are used to place attributes of a variable into some kind of order. This ordering can be from the highest to the lowest, top to bottom, largest to smallest, best to worse and there are many other similar possibilities. This type of scale can be used to gain respondents' perception of, for example, quality of a product or service. This information can be very useful on its own and can also be compared with demographic information on the questionnaire survey as part of the analysis process. However, there is a key issue with ordinal scales. This relates to the gaps or intervals between successive points on the scale. You may believe, if you ask a visitor to rate their dining experience of three different restaurants on an ordinal scale, from best to worst, that if they rated one restaurant as the best giving it a '1' and then placed another as second with a '2' and then the third with a '3', that the sizes of the gaps or intervals between '1' and '2' and '2' and '3' are identical. However, this may not be the case at all. The visitor may, in fact, believe that the restaurant they have indicated as the best and given it position '1' on the ordinal scale is far better than either of the other two restaurants. Additionally, the visitor may have decided the restaurant that they put as '2' on the scale was only marginally better than the restaurant to which they gave the '3'. Therefore, using this type of ordinal scale will not put you in a position to know how the respondents interpreted the intervals on the scale.

Interval scales

The third type is the interval scale. This type of scale is used commonly in academic research in the social sciences and is appropriate for questionnaire survey work in tourism and related fields, and it overcomes the problem of 'interval size' on the ordinal scale. Here, the intervals between successive points on the scale are defined and are usually equal. It is common for the interval between each successive point to be one unit. The interval points on such scales can range from three, up to nine or even more than this, however five- or seven-point scales are particularly common. Likert scales and semantic differential scales are of the interval scale type, and these are discussed in greater detail in the section following this general discussion of scales. With its defined interval, this type of scale clearly has an important advantage over the ordinal scale. One key aspect of the scale having defined intervals is that it is possible to convert the scale directly to numeric values, as will be discussed in the section on Likert scales and semantic differential scales.

Ratio scales

The fourth type is the ratio scale. Ratio scales are similar to interval scales and ordinal scales but are more open-ended. Although they usually start at zero, they can go up to a very high figure and theoretically could go up to infinity. Ratio scales are used to collect ungrouped or unclassified data. For example, you may want to ask a hotel chain how many staff it employs. You have decided to separate the categories into full-time and part-time employees. However, you do not know precisely how many full-time staff or part-time staff are likely to be employed, but you do know that there are many more part-time staff overall. It is possible in your survey, that one hotel has twelve full-time staff and forty-three part-time staff during one year, while another has twenty-one full-time staff and seventy-six part-time staff during this same year. Therefore, using just two hotels you will need to have one category of full-time staff, ranging from 12 to 21 and a category for part-time staff between 43 and 76. Therefore, there will be a problem when deciding the dimensions of your categories, and even more so when other hotels are included. However, the general principal to be aware of is that, under these circumstances, it will be difficult to create equal interval categories on your survey, so a ratio scale would be suitable.

Measuring attitudes

One of the key areas researched in tourism and related subjects using a questionnaire is attitudes. This can be, for example, attitudes to service in a hotel or restaurant, or to quality of a leisure centre's facilities, or attitudes of visitors to their experience at a theme park. But what exactly are attitudes and what are key issues in trying to assess them?

Riley (1996: 75) defined an attitude in the following way:

> 'A predisposed response to situations, objects, people, other self-defined areas of life. It has both a perceptual element and an affective component. The latter produces a direction in the attitude – positive or negative. This in turn can influence the perceptual element – we see what we want to see.'

As Clark et al (1998) state our attitudes are based on our experience and are organised in a particular way. Clark et al also indicate that attitudes are focused, closely related to our emotions and likely to be 'fixed'. By fixed, they mean that attitudes are usually deeply held views that are linked closely to our values, our culture, the behaviour we usually adopt and also what we regard as appropriate behaviour in specific circumstances.

There are certain characteristics of attitudes, one of which is in the definition of Riley (1996) above, and these are summarised below:

- They are related to an object or person or an idea;
- They influence the way we perceive things and therefore how we collect information on things;
- They are learned and are long lasting;
- They imply evaluation of things and, related to this, the way we feel about things.

(Clarke et al., 1998).

Clark et al (1998) state that there are three components of an attitude. The first of these they term the *cognitive*, and this refers to how we are aware of, and understand, an object. Certain words are associated with this aspect of an attitude and are to do with the origin and location and these words include terms such as 'comes from', 'produces', 'results in' and 'causes'. The second component of an attitude, according to Clark et al., is *behavioural*. This is concerned with action that is implied by the perception of the object. Typical words associated with this, all of which are verbs, include 'choose', 'hire', 'fire', 'reject', 'buy', 'sell'. The third component is what Clark et al term the *affective*. This is to do with our evaluation of and feelings towards the object. Typical words connected with this affective component include 'like', 'dislike', 'happy' 'sad', 'love', 'hate' and 'bored'.

Our attitudes are formed and developed over time – we are not born with them. Our own experience contributes greatly to this formation, but the culture in which we live, the influence of others and the general views in society are also important in attitude creation and development. Our attitudes are also conditioned by whether or not they are accepted by our family, friends and close colleagues. So, our attitudes are very important to us as individuals and perform certain functions for us such as to help define precisely who we are, to move us away from people we don't like to those we do, and to help us to know how to react and behave in new circumstances that are similar to ones we have experienced in the past.

Attitudes are also important because they have a strong influence on the way we behave. Although it is not necessarily possible to predict behaviour precisely based on an individual's attitude, it is clear that this link does exist. You may have asked yourself the question: 'Was my behaviour to do with the attitudes I have?' You may also have thought, even if not in these precise words: 'Is my attitude an 'after the event' justification for my behaviour?' These questions indicate how closely linked attitudes and behaviour actually are. If an individual repeatedly behaves in the same way to similar circumstances that occur over time, they will inevitably take on the attitude that goes with the behaviour. Conversely, if we have an attitude linked closely to a personal value when confronted with the need to respond to a particular event or stimulus, then we are very likely to follow our attitude and behave accordingly.

As Clark et al (1998) indicate, our attitudes are often grouped and are organised in an attempt to create a consistent view of the world, enabling us to behave accordingly. Clark et al also state that some attitudes are subsumed with others. So, for example, an attitude statement 'I love cricket', may be associated with 'physical fitness', and be subsumed within a larger attitude of 'I like team sports' or, 'I like competition'. The attitude 'I hate cricket' could be associated with a larger attitude 'I prefer reading to physical activity', or 'I do not like competitive team games' or 'I prefer individual sports activities to team games'. So there are many ways in which attitudes can be interpreted and attitude measurement is a way of finding out the nature of various attitudes and the relationships between them. In tourism and related fields, it is relatively easy to think of themes and topics where attitudes are important. For example, when considering tourists, which destination is chosen for a summer break, the cost of the trip, the amount of money to be spent on holiday, which restaurants to be used, which cafes and bars will be visited, and which types of entertainment will the tourist gets involved in will all be based on their values and attitudes.

Therefore, in terms of research, attitudes can be very revealing about respondents and questions about attitudes are an important part of most questionnaire-based research. The next section discusses scales for measuring attitudes as well as problems associated with this measuring process.

Likert scales

Both Likert scales and semantic differential scales can be used to record respondents' attitudes to a particular topic, theme or issue. For example, a very common type of Likert scale is shown below and would be used to gain respondents views on, or attitudes to, a statement on a particular topic.

Strongly disagree	Disagree	Neither agree nor disagree	Agree	Strongly agree

This Likert scale has five points and the interval between each point is intended to be equal, although as indicated earlier in this chapter, there are problems with assuming the intervals between the values are equal. On this scale, strengths of view can be indicated and there is a mid-point which is, in effect, a neutral position, and to one side of this there are two equally spaced intervals of 'negative' statements, one a stronger view than the other and on the other side are two equally spaced intervals of 'positive' statements, one a stronger view than the other. As the intervals are equal on this Likert scale, the responses can be converted to numeric values as shown below.

Strongly disagree	Disagree	Neither agree nor disagree	Agree	Strongly Agree
1	2	3	4	5

It is then a relatively easy task, when conducting your analysis, to manipulate these numeric values and produce simple, but very useful statistical values, such as the overall mean (average) score for a particular statement. The mean score for one statement can then be compared with the mean for other statements that have used a similar Likert scale on the questionnaire.

There are several important issues with the use of Likert scales. One is not to do with the scale itself but is concerned with creating a question that says precisely what you want it to say. Pre-testing questions or using a pilot study may be the only ways that you can feel confident that respondents interpret your questions in the way you intended them to be understood.

Presentation of statements

Another issue with Likert scales is how to present the statements. As indicated above the statements are presented individually, but should be related and focused around the particular topic. In fact, when using the Likert scale, it is usually the case that you would link, via a question number, e.g. Question 7 and then list as Question 7a), 7b), 7c and so on. Doing this informs the respondent that the statements are linked. They can also be informed about this in the instructions on how to answer using the Likert scale response category statements in the question. Note that you do not have to present the response categories from 'strongly disagree' to 'strongly agree', with the respective numbers underneath – the words of categories are sufficient. However, although it is common to use a five-point scale, it is also possible to use seven-point and nine-point scales. In each of these cases, it is conventional to just indicate each of the opposite ends of the scale.

An example of how you would present a nine-point scale is shown below.

Strongly disagree	1	2	3	4	5	6	7	8	9	Strongly agree

Nevertheless, an important question in relation to the points on the scale, is precisely how many should you use? The guiding principle is that if you present too few, it is difficult for the respondent to differentiate and represent accurately the distinctions, and they may feel frustrated. However, if there are too many points on the scale, the respondent may just make an arbitrary choice. Experience suggests that five really is the minimum, and probably nine is the maximum number of points on a Likert scale.

When presenting the statements used for a Likert scale to respondents, it is important that you try to prevent respondents slipping into what could be termed 'automatic response mode'. This means that if there are ten statements to respond to and the respondent has ticked 'agree' to the first three, what you want to avoid is the respondent merely ticking all ten with 'agree' in the same position on the respective scale for each statement. This means that either you will have to reverse the order of the Likert scale running from left to right, so that instead of going from 'strongly disagree' to 'strongly agree' it goes from 'strongly agree' to 'strongly disagree', or you will need to present a negative statement rather than a positive one to respond to. The following example indicates what is meant here. If you were investigating the social impacts of a music festival and the Likert scale statements were mainly positive in terms of, for example, enjoying watching bands performing, enjoying socialising with your friends and family, gaining personal fulfilment, and learning a new skill, you could include a 'negative' category such as this statement 'The music was too loud', or 'The audience behaved badly'. Clearly respondents indicating that they agree with these statements are indicating negative aspects of the festival. However, if your respondents have indicated they agree with the positive statements then you would expect most of them to disagree with the more negative statements. So, presenting some negative statements is an attempt to prevent your respondents casually ticking the same boxes, in the same place and, instead, making them read your statements carefully and hopefully getting them to answer accurately. Hence, this approach is also an attempt to make your results more reliable.

The importance of the mid-point

One thing you have probably picked up on already, is that all the Likert scales have an odd number of points. Why is this? The reason is that there can be a mid-point, with an equal number of 'positive' statements to one side and the same number of 'negative' statements on the other side. However, what precisely does this mid-point signify? Strictly speaking it means 'neither agree, nor disagree'. However, it is possible that respondents could interpret this mid-point as 'Don't Know', which could mean 'Don't Care'. But this is not quite the same as 'Neither Agree, nor Disagree', although it may seem so at first glance. An important difference between them is that 'Don't Know' could mean 'I have not made my mind up yet', or it could mean 'I don't care about answering this'. But it is also possible to interpret the response 'Neither Agree, nor Disagree', as 'I am not interested in this topic', or 'I have given it thought and I do not agree with the statement, but neither do I disagree with it'. This may seem a relatively fine difference, but it is important, and many users of Likert scale questions provide a separate category outside the scale, which is for 'Don't Know'. One reason that the 'Neither Agree, nor Disagree' response is important is that it is at the mid-point of a scale. Therefore, on a five-point scale the numeric value is a '3'. However, the 'Don't Know' position, being outside the scale, would have a numeric value of a '0'.

Now, you may be surprised to know that the mid-point position on the Likert scale and what it actually means has been the subject of much research (see Ryan, 1995). Over and above the discussion that has been presented earlier on whether it is interpreted by respondents as 'Neither Agree, nor Disagree' or 'Don't Know', there is another issue. Experience of many years of using the Likert scale in the social sciences has indicated that a very large number of respondents to Likert scale questions give answers close to the mid-point. This leads to many mean scores for Likert scale statements being between 2.5 and 3.5. This 'clustering around the mid-point' is therefore common. However, given that the scale is attempting to measure strength of attitudes, clustering around the mid-point suggests that respondents do not like to give strong views! Couple this with the issue of how, precisely, the mid-point is interpreted by respondents, and it should be clear that Likert scale questions have this 'mid-point issue' as an important limitation.

The issue of 'clustering around the mid-point' also raises, even if indirectly, the problem of how individuals respond to any sort of Likert scale. We assume when putting together the statements for the scale that there is an equal interval between each successive value 1, 2 etc. up to 5, or 7, depending on the scale. But how do we know whether respondents react in this way? How do we know if Respondent X, who answers by circling a 2 on a 7-point scale, means the same as Respondent Y who also gives a 2 on this scale? The simple answer is, we do not. However, we make the assumption that each respondent is interpreting the scale in the same way as every other respondent. To help indicate whether this is a valid assumption, a pilot survey and/or pre-testing can help. However, even after a 'successful' pilot survey, we can still not be certain of the way our sample is responding, but we should be in a position to at least feel more confident.

Although the Likert scale is attempting to measure the strength of feeling in relation to a statement, it is often the case, during analysis, and this is done on occasions in relation to the problem discussed in the previous paragraph, that the categories 'Strongly Agree' and 'Agree' are combined. This gives an aggregate 'positive' total which may be expressed as a percentage, and likewise the 'Strongly Disagree' and 'Disagree' categories are combined to give an aggregate 'negative' total. Note, that the key rationale for doing this is so that the overall percentage of respondents who 'agreed' (responded positively to the statement) can be compared with the overall percentage of respondents who 'disagreed' (responded negatively). Aggregating the 'positives' and 'negatives' in this way provides an opportunity to compare percentages with other statistics for a Likert scale statement, particularly the mean or mode. It may be the case that 64% of the sample have 'agreed' (the aggregate value of those responding: 'Strongly Agree' and 'Agree' with a statement that has a mean of 3.9. So, these two figures support each other. In another example, a statement could have 61% who 'disagreed' (the aggregate value of 'Strongly Disagreed' and 'Disagree') with a mean of 2.2. So, once again, these figures support each other.

Bi-polar semantic differential scales

This type of scale is similar to the Likert scale. However, unlike Likert scales, at each end of the semantic differential (the ends of the scale are the 'polar' part of the scale and as there are two ends, it is a 'bi-polar' scale) is usually an adjective. The adjectives are opposites, for example: 'good' and 'bad', 'wet' and 'dry', 'comfortable' and 'uncomfortable', 'clean' and 'dirty', 'friendly' and 'unfriendly'. The scale runs between the two opposite ends of the scale. The intervals on the scale are usually seven equal points, or sometimes nine or even more. The assumption in a semantic differential scale is that the semantic dimension conforms to an attitude (Clark et al., 1998). When preparing the words for use in the semantic scale, care must be taken to select terms that are appropriate. For example, if you were involved in research into motivations for desert-based holidays and were interested in the importance of climate, it would probably not be a good idea to use the terms 'wet' and 'dry' at opposite ends of your scale, as 'wet' would be largely redundant in this context.

Below is what a seven point bi-polar semantic differential scale could like on a survey.

Good 1 2 3 4 5 6 7 Bad

The rationale for the number of points, or gradations, on the semantic differential scale is very similar to that of a Likert scale. Too few points and respondents will feel that they cannot accurately represent the distinctions, too many points and respondents will feel that they are not be able to accurately distinguish between, for example points 8 and 9 on an 11-point scale. However, some users of the scale argue that it is not necessary to include a numbering system (see Clark et al., 1998). Nevertheless, without an actual numbering system, it will not be possible to conduct statistical analysis on a bipolar scale.

There are yet others users of semantic differential scales who number the scale from the midpoint (as shown below). Here there are seven points on the scale with values from 1 up to 3 for 'Good' and to the right of the mid-point 0, from 1 to 3 for 'Bad'. This numbering system makes it very clear to respondents that '0' is 'certainly the mid-point and neither good nor bad'.

Good 3 2 1 0 1 2 3 Bad

Semantic differential scale questions, with numbering, have the same advantages as Likert scale questions, in that they can be easily analysed as they are in numeric form. However, as there are only two words – the adjectives with opposite meanings at each end of the scale – it is less clear whether all respondents have exactly the same understanding of, for example a '6' on a '7' point scale, or a '2' near the end with the adjectives 'Bad' when using the scale with '0' as a mid-point. On a Likert scale with a five-point scale, from 'Strongly Disagree' to 'Strongly Agree', each number 1 to 5 has a specific word or phrase attached to it, this is clearly not the case with a bi-polar semantic differential scale.

Semantic differential scales also have the same 'responses clustering around the mid-point' issue that Likert scale questions exhibit. One solution to this is that a scale with an even number of points e.g. six or eight could be used. However, if respondents are familiar with a seven-point or a nine-point scale, they may be confused and/or annoyed by a scale with an even number of points. In a similar way to Likert scales, it may also be necessary to have some form of a 'Don't Know' category, which is not the mid-point of the semantic differential scale.

Likert scale questions have tended to be used in tourism and related studies more than bi-polar semantic differential scales. However, both have their uses and can be relevant to the topic you are investigating. Whether using one or the other, an important factor is to ensure you provide respondents with clear instructions. Therefore, if you are using a Likert scale, you should put instructions such as:

> Please indicate the extent to which you agree or disagree with the following statements by putting a ring around the appropriate answer.

If you are using a bi-polar semantic differential scale, you should put instructions such as:

> Please indicate the appropriate number on the scale by putting a ring around it.

Checklists

Yet another type of closed-ended question is the checklist. This can be used if attempting to find out, for example, if respondents play a number of sports, have taken holidays in certain destinations or have eaten at specific types of restaurants. In each of these cases, the checklist would be in the form of a box with a list of factors that can be ticked. With checklists, unlike several other types of closed-ended question, it is possible that more than one category can be ticked.

A possible problem with checklists is that you may be tempted to produce a very long list, to get as much data as possible. However, this can be counter-productive, as respondents may lose interest before completing the checklist. On the other hand, a checklist may prompt some respondents to tick more categories than if an open-ended question was asked and respondents had to think and put a response in their own words. For example, respondents may be asked which sports they take part in regularly and some respondents may tick far more sports than is actually the case, possibly because they feel they should be playing those sports, even if they are not. So, checklists of this type can inflate responses.

Rankings

You may wish to ask respondents to put certain factors in some kind of rank order. This could be linked to their preferences to, for example, a number of hotels or restaurants. Let us assume that your study is of five Chinese restaurants in one area, and you want your respondents to put the restaurants in an order from 1st to 5th on the basis of food quality. However, be aware that although you may think this is a good strategy, it may be difficult for respondents. If you have asked them to rank the restaurants according to food quality, this may be difficult for some respondents as they find they are unable to rank on just food quality, as for example price is also very important to them. For some other respondents, service may be an important ranking criterion. As with ordinal scales, there is also the issue of the interval on the scale between the five restaurants – some respondents may have just one restaurant they rate highly and put it in rank position '1' and feel the other four restaurants are fairly equal – they could rank these as '2' but may feel that they should all be ranked '5'. As a result of these issues, rankings can also be difficult to analyse (Long, 2007).

Preferred statements

On a questionnaire, these may look rather like multiple choice questions. They can be difficult to construct, but are often used to represent different positions in a political debate. Consider the example below, which I used in research at a festival.

What is your view on the number of visitors to the Festival? Please tick one box only.

- ☐ There are too many visitors to the Festival.
- ☐ There are just about the right number of visitors to the Festival.
- ☐ There are not, as yet, too many visitors to the Festival.
- ☐ I don't know.

An issue with preferred statements is that if respondents have recently completed a multiple-choice task, separate from your research, they may feel that there is one 'correct' answer. Clearly this is not the case in the example I have given above, as respondents are being asked to give their opinion.

Types of questions to avoid

The previous discussion has revealed that there is a large range of possible questions that can be included on a questionnaire, but there are a number that you should try to avoid. One important type of question to avoid using is the leading question. This is the type of question that leads respondents to give a specific answer. Here is an example 'Women are not as good as men at sport, are they?' Hopefully, this is not a question you would use on a questionnaire. Other than it will probably be seen as insulting by a potentially large number of respondents, you are leading the respondents by the way you have constructed the question. They are likely to interpret this question as indicating that they should respond 'Yes'. Let us assume that you would not use a question such as the one above. It is more likely with the best of motives you use a question such as: 'Why are women not very interested in mountain biking?' The reason for this question (your main motivation), let us assume is, that you want to ask a follow-up question: 'What needs to be done to encourage more women to take part in mountain biking?' But, with your first question, you have set up the respondents to feel that they have to agree with you that women are not interested in mountain biking. However, they may not actually believe this. In this particular example, it may be sufficient just to ask the question, which is currently represented as the follow-up above, and leave out the leading question.

Another type of question to avoid is the hypothetical question. This is the type of question in which you have the phrase: 'What if…? Or just 'If…?' For example, you may be interested in the relationship between the entry price to an attraction and motivation to visit it. You may even have a hypothesis which states, something to the effect that, 'an increase in entrance charges will lead to a decline in number of visits'. Now using this hypothesis as your guide, you could ask the question: 'If the entrance price was increased by 25%, how less likely would you be to come to the attraction?' If you think about this for a moment, it is difficult to give an accurate answer. How can you put a figure on the 'less likely to' part of the question? A respondent might respond '25% less likely', but what does this mean? Does it mean, if they had planned to visit four times, they will now only visit three times? The respondent may actually think: 'I have visited once, I have no desire to visit again, whatever the price', but they will not be able to provide this response to the question you have asked. Another thing to note here is that this is actually a leading question, as you have assumed that an increase in the entrance price will lead to a decrease in the number of visits to the attraction! A better question on this theme, in this case it is an open-ended one, would be to ask: 'What effect would an increase of 25% in the entrance charge have on your motivation to visit the attraction?' This question provides an opportunity for respondents to give either a detailed response relating to the effect, or possibly state 'No effect'.

Yet another type of question to avoid is one that is too complex and possibly ambiguous. Let us assume that you are interested in the rationale for tourists' choice of particular holiday destinations and you might be tempted to ask the following question: 'What is the most important factor influencing your holiday choice: the price of the holiday; the length of the journey to the destination; the type of transport to the destination; the quality of the accommodation; the average daily hours of sunshine; whether local people speak English?' All of the factors would appear to influence holiday destination choice. But this is clearly a complex question and you have not made it clear whether you want your respondent to give just one answer from the list, more than one, or possibly refer to all and put them in descending order of importance. If you wanted to ask such a question it would be better to have checklist or ask respondents to rank the factors.

Question sequencing

You may have made a number of decisions on the types of questions you wish to use on your questionnaire, but a very important factor relates to the order or sequencing of the questions. You need to consider how you can put the questions in a sequence that makes it both easy for respondents to answer, but also for you to analyse. There may seem to be no obvious order when you look initially at the questions you have designed. However, there are ways to set out your questionnaire to make it easy to respond to, but also that give useable responses. According to Veal (2011) questionnaires are commonly divided into three sections: respondent characteristics, activities and behaviour, and attitudes and motivations. Nevertheless, the actual sequencing is open to debate. Some researchers, including Veal (2011) and Long (2007), suggest strongly that the demographic information (respondents' characteristics) should be placed at the end of the survey. One reason given is that if respondents are becoming weary towards the end of the questionnaire, they will find giving demographic information an easy task (Long 2007). Other researchers (see Mason, 2001; 2003) indicate that it may be preferable to place the demographic questions at the beginning. Veal (2011) also argues that the confidence of the respondents should be obtained before personal questions are asked, so the demographic information should be at the end. However, Mason and Cheyne (2000) argue that placing the demographic information at the beginning is appropriate, as it is easy for respondents to give this, which will help them 'warm-up', and also other types of question, for example, attitude ones will require more thought. Hence, Mason and Cheyne (2000) argue the demographic information should appear first on the survey and questions specific to the main topic should follow these. In addition to preferring to have demographic questions at the beginning because for respondents completing information about themselves is straightforward, Mason and Cheyne (2000) claim that it is less likely respondents will fail to complete this section, if it is at the beginning.

This discussion has indicated that advice on sequencing varies, and it is largely up to you to select and provide a rationale for the order of your questions. However, I tend to employ the following approach, depending on the topic and types of questions I want to use.

1 After the initial demographic questions, I will use closed-ended questions, such as 'Yes/No' type. Such closed-ended questions should be relatively easy to respond to.

2 I follow these with other closed-ended questions, such as those that require ticking a checklist, or filling in a table with one or more answers.

3 I then have 'attitude' questions – those using an attitude scale, such as Likert scale questions; these are also types of closed-ended questions. Attitude scale questions require more time and concentration from respondents. Such questions are very likely to be particularly important in their own right, but also are to be used in connection with demographic information, so they occupy a position in the sequencing of questions which is at a mid-point, when respondents should be 'warmed up', but still fresh and focused on the task of answering the questions.

4 Towards the end of the questionnaire, I use open-ended questions, if I am using any of this type of question.

5 Finally, the last questions are likely to be short, possibly closed-ended but may be open-ended, asking respondents to add comments on topics/themes that they feel could have been included in the questionnaire, but have not been.

Sensitive questions

Some questions will be regarded by a number of people as containing sensitive topics, other questions will be recognised by almost all respondents as sensitive and there may be a few people who regard some topics as problematic, yet you may have thought of these as being not in the least bit sensitive. Giving your age to a stranger is not something all respondents will do willingly – this is why it is better to have age-band categories ranging over (at least) a decade. Income levels can be a very sensitive topic. Again, as with age, it is better to have categories with a band range of, say £5000 or £10,000 rather than asking a direct question about income.

Clearly, sensitivity varies from individual to individual, as well as in relation to culture, so although it is difficult to generalise, some nationalities are more likely to answer certain questions than others and you need to be aware of this before you embark on a topic that is potentially sensitive with your respondents. A way to reveal if your questions are insensitive is to pre-test them and then as is likely, conduct a pilot questionnaire survey.

The problem with possible sensitive questions is a reason that some researchers give for putting the demographic questions at the end of the questionnaire, rather than at the beginning. As indicated earlier, I usually put demographic questions at the beginning (for the reasons provided above), and because of this I am very careful in my attempts to avoid sensitive demographic questions.

However, depending on the topic, it is quite possible that there could be sensitive questions in the main section of your questionnaire. As Case Study 4.1 indicates, disability is a sensitive topic, and needs to be dealt with very carefully. Other topics that you may want to ask about which are very likely to be viewed as sensitive by respondents include gambling, alcohol consumption and aspects of sexual behaviour. Before you ask a question, you should ask yourself: 'Is this a sensitive topic?' If you believe the answer is 'Yes', then you should proceed very carefully with your question construction. You should also ask yourself: 'Is this question necessary?', 'How does it relate to my aims and objectives?' 'What do I hope I will find out by posing this question?' If you cannot provide convincing answers to these related questions, then there is little point in asking the question. Remember, you do not want to put off your respondents from completing your questionnaire in full, because they are unhappy about one insensitive question

Case Study 4.1 presents a draft survey that a student prepared and sent to me for comment well in advance of using it in the field. The questionnaire was intended to be used at three different heritage attractions. The intended respondents were visitors to each of the locations. The focus of the student's research was visitor attitudes to each of the sites and related management aspects. The case study shows the student's first draft of the questionnaire survey, with my added comments, as supervisor. Some details have been changed to maintain the anonymity of the student. Also, the amount of space for respondent comment on the questionnaire has been reduced for publishing purposes.

Case Study 4.1: Student designed questionnaire: first draft

First draft of a student's questionnaire with the supervisor's suggestions.

(Supervisor's comments in italics)

1 Have you ever visited one or more of the following heritage sites?

A UK Cathedral	
A major UK heritage attraction	
A major heritage site in Eastern Europe	

How do respondents answer? Yes or No?

2 Did you know about the sites prior to your visit?

Yes	
No	

This is too complex, as there are three possible sites in question 1), so which one should the respondent consider? Also, largely a redundant question, unless the respondent was forced to go to the site!! Better to write. 'What did you know about the site, if anything?' or in the context of what you want to gather information about here, 'how did you find out about the site?'

3 Was there enough information available?

Yes	
No	

This question is OK. But what do you learn from it? A good idea to add: 'Please explain your answer'

4 Was the heritage site(s) the main purpose of visiting the city?

How do you respond to the question? Yes/No? The box here looks as if the respondent should give a reason as well. Consider, is this necessary?

5 Were the sites that you visited well maintained/run-down according to you?

Poorly maintained	
Somewhat maintained	
Well-maintained	

It is good to have a range of possible response categories, but why three? It appears that this is an attempt to have a mid-point category with a 'negative' category on one side, and a 'positive' category on the other side of the mid-point. But it is not clear at present with 'somewhat maintained' as the mid-point. It could be better to have the following: 'very poorly maintained', 'poorly maintained', 'neither maintained well or poorly', 'well maintained', 'very well maintained'. There are now five categories, a clear mid-point (neither one, nor the other) and two possible response categories either side of the mid-point. This is now a form of Likert-scale question.

6 What were your expectations of the site and were they met? (Please, specify in the box)

This topic is important for the research but it is two questions in one. It is probably better to have the following:

5a) What were you expectations of the site? Please respond in the box – this is then an open-ended question.

5b) Were you expectations met? Yes/No. And possibly have an open-ended part to the question, such as: 'explain your response'.

7 Was there sufficient information/signage provided?

Yes	
No	
Unsure	

An important topic, but again this is two questions in one. You should have a separate question on each of 'information' and 'signage'. Also how do respondents answer – with a tick in the appropriate box? You need instructions to tell them.

8 How would you rate the quality of the information provided?

Very Poor	
Poor	
Average	
Good	
Very Good	

4

I understand what you are trying to do here, but what does average mean? I believe you intend it to mean, 'neither poor, nor good' but it does not mean that. You should have a category in the middle 'neither poor nor good'. Also you need an instruction 'please tick the appropriate box'.

9 Were there enough facilities provided (e.g. toilets, cafes, playgrounds)?

Yes	
No	
Unsure	

This is OK, but is 'unsure' meant to mean 'do not know'? Once again you need instructions on how to complete – 'please tick in the appropriate box'. Do you want respondents to explain their answers?

10 Do you have any health problem and disability that limits your daily activities or work?

Yes	
No	

Be aware that this is a sensitive topic. Consider ethical issues. It is also complex and confusing. 'health problems' are not necessarily the same as 'disabilities' - many people are short sighted (a health problem), but probably do not regard themselves as 'disabled'. Will you have a comment about this in your covering letter? How important is it to include this in your research?

If NO, go to question 16.

Are you sure this is correct? It appears that questions 11-15 are relevant as well.

If YES, how would you rate the facilities for people with special needs on the site?

Very Poor	
Poor	
Average	
Good	
Very Good	

Once again what does your category 'average' mean here?

11 Are there any other facilities that you would like to see or existing ones you would like to see improved? (Please, specify in the box)

This is really two questions. Subdivide into two parts – a) other facilities, b) improve existing facilities

12 What do you think would improve the appearance of the place? (Please, specify)

This question is vague – it is not precise enough. What does 'appearance' mean precisely and what do you mean by 'place'?

13 How likely are you to return?

Very unlikely	
Unlikely	
Unsure	
Likely	
Very likely	

Five categories is a good range of possible answers, but once again what does 'unsure' mean? Don't know? Don't care?

14 How well is the site interpreted/animated for tourists?

Poorly interpreted	
Somewhat interpreted	
Well interpreted	
Excellent interpretation	

This appears to be two questions. I suggest that you drop the 'animated'. This is not a term often used in tourism/visitor management, so it could be confusing. You have four categories. It would probably be better to have five. I suggest remove the 'somewhat interpreted' as this is too imprecise and use terms ranging from 'very poorly interpreted' to 'very well interpreted'. Make the mid-point, 'neither poorly, nor well interpreted'.

15 How would you rate the level of your experience?

Very Poor	
Poor	
Average	
Good	
Very Good	

Your response categories are generally OK. I would prefer it if you used 'neither good nor bad' as your middle category. Do you want respondents to explain their answer?

16 Describe any specific experiences/feelings while visiting the site(s)?

Once again this is really two questions a) experiences b) feelings. What do you want to get from this question, what is the aim? Do you want one 'key' reaction or…?

4

As Case Study 4.1 indicates the student has used a range of different type of questions, most of which are closed-ended. However, my comments show you the type of errors that are quite common, particularly in a first draft of a questionnaire. These include lack of clarity, two questions combined into one, problems with Likert-scale type questions in terms of not sufficient response categories, lack of clear instructions and the use of a sensitive question (which is likely to be unnecessary in the context of the research aims and objectives). These errors could lead to possible confusion amongst respondents and the likelihood that they will not answer questions accurately, meaning the student's results will not be reliable. Attempting to analyse the responses will be difficult. The wider effect of these combined issues will be to reduce the overall quality of the study.

Examples of questions

■ Closed-ended questions

Below I present and discuss a selection of closed-ended questions I have used in my own research. Figure 4.1 shows a specific type of question that I used in research I conducted at a zoo in New Zealand. The main aim of the research was to investigate the roles of the zoo. The four statements were used in a Likert scale question which had the conventional five response categories ranging from 'Strongly Disagree' to 'Strongly Agree' for each statement.

Figure 4.1: An example of Likert scale statements used for research at Wellington Zoo, New Zealand

It is an organisation that mainly breeds rare animals
It is an organisation set up mainly to educate people
It is an organisation set up mainly as a tourist attraction
It is an organisation set up mainly to conserve animals

Notice in each statement in Figure 4.1, at the beginning, there is very similar wording, but the end of each statement contains the issue that is different from the other statements. This use of statements with (partly) identical wording was done in an attempt to ensure that respondents were comparing like-with-like and possibly would not be confused, which could have occurred if the wording had been different. The statements produced interesting, useful results including a very high percentage of respondents in agreement with the statement concerning the educational role of the zoo and a small majority who disagreed with the statement concerning the zoo being mainly a tourist attraction. These results suggest the statements were valid in relation to the topic and aim of the research project.

I also used some checklist questions in my questionnaire at Wellington Zoo. One of these, shown in Figure 4.2, had discrete categories and respondents could only tick one category.

Figure 4.2: An example of a checklist question: Motivations for Visiting Wellington Zoo

Family outing	
To see the animals	
A day out	
School/other group	
Visit with a friend	
Other	

The intention of this question was to reveal the main motivation for the Zoo visit. The categories of response were derived from literature on motivations for zoo attendance generally, previous surveys at Wellington Zoo and the advice of the Zoo management team. Be aware that the 'Other' category has been used; this is an attempt to ensure that, if there is any category that has been missed, respondents can tick this one. Note, also in this checklist the category 'School/other group'. Strictly speaking this is not a 'motivation' category, but discussions with zoo managers had indicated that there was a relatively large number of zoo visitors who came in groups, and I should include this category in the question. Therefore, an important point to be aware of here is that it is always worth considering authoritative advice. In this checklist question, when completed, the percentages should add up to 100%. Responses to this question are shown in Figure 7.1 in Chapter 7, which discusses the presentation of results.

I used another type of checklist question which allowed respondents to tick more than one category. This presented statements for respondents to tick in relation to the specific roles of the zoo. This is shown Figure 4.3.

Figure 4.3: An example of a checklist question: Wellington Zoo roles

Role	
A place of relaxation?	71%
A place for education?	79%
A place where visitors can see animals entertaining them?	46%
A place for research?	37%
A place for conservation?	55%
Other	6%

As respondents could tick as many categories as they wished, totals do not add up to 100% (this would only occur if respondents could tick just one category). Figure 4.3 indicates the actual percentage of total respondents to the questionnaire who ticked that category. It is easy to see the importance of 'education' and 'relaxation' as zoo roles from this question's results. Note also that the 'Other' category has only a small percentage of responses, indicating that I had not omitted an important 'role' category in the question categories, but still would enable respondents to answer in another category than those in the box.

An example of a closed-ended question with one choice only is shown in Figure 4.4. Note the instructions indicate that only one choice can be made and that there is no overlap between the four response categories. Also note that the intervals between each of the response categories are not the same.

Figure 4.4: An example of a closed-ended question with one choice only.

Please indicate below whether or not you have visited Stonehenge before (please circle one only):

First visit	Once before
2-5 times	More than 5 times

Attitude questions

I now present and discuss a number of attitude questions that I have used in my research. First, is a Likert scale question which was used in my research at the World Heritage Site of Stonehenge. I conducted the research via a face–to-face questionnaire at the exit of the turnstile of the old visitor centre. This was replaced by a new centre in late 2013/14.

Figure 4.5 does not include all the statements used in the actual question, but provides a selection of them. Note that a five point Likert scale has been used. Also note the instructions on how to complete this question.

Another example of a Likert scale question is shown in Figure 4.6. This is just a selection of the statements used in the question and does not include all of them. Note that some of the statements are 'positive' and some are 'negative' in terms of the topic of 'views of respondents on their visit to Stonehenge'. Statements d), e) and j) are 'negative' statements. These three statements are an attempt to ensure respondents read the questions carefully and do not just run through the list as quickly as they can, ticking each with the same response category

Figure 4.5: An example of a Likert Scale question

Motivations for Visiting Stonehenge

Please provide your reasons/motivations for visiting Stonehenge below. Use the 5 point scale which ranges from 5 (Strongly Agree) to 1 (Strongly Disagree), where:

(5) Strongly Agree (4) Agree (3) Don't Know (2) Disagree (1) Strongly Disagree

Please provide a response to all statements.

a) A day trip away from home	5	4	3	2	1
b) I am interested in prehistoric monuments	5	4	3	2	1
c) To expand my knowledge	5	4	3	2	1
d) Stonehenge is unique	5	4	3	2	1
e) To be with my partner/family	5	4	3	2	1
f) Stonehenge has a spiritual value	5	4	3	2	1
g) To be with friends	5	4	3	2	1
h) To escape from routine	5	4	3	2	1
i) Stonehenge has a religious meaning	5	4	3	2	1
j) It is part of a tour of major UK tourist sites	5	4	3	2	1

Figure 4.6: Views on visiting Stonehenge

Please consider the following statements and use the 5-point scale to indicate your level of agreement. Once again, the scale is as follows:

(5) Strongly Agree, (4) Agree, (3) Don't Know, (2) Disagree, (1) Strongly Disagree

a) Stonehenge makes an important contribution to our understanding of the past	5	4	3	2	1
b) Signs and signage are clear and easy to understand	5	4	3	2	1
c) The hand held audio-guides (wands) are useful	5	4	3	2	1
d) There is too much noise from the roads near to Stonehenge	5	4	3	2	1
e) Stonehenge is overcrowded with visitors	5	4	3	2	1
f) Visitors to Stonehenge behave well	5	4	3	2	1
g) Not allowing visitors to touch the stones is necessary to conserve Stonehenge	5	4	3	2	1
h) The toilet facilities are good	5	4	3	2	1
i) The gift shop is good value for money	5	4	3	2	1
j) Parking is a major problem at Stonehenge	5	4	3	2	1

Figure 4.7 is an example of the use of a Likert scale where results have been aggregated. The research from which this information has been extracted involved local residents' attitudes to the proposed setting up of a small tourism development – a cafe/bar – in a relatively isolated part of rural New Zealand, where there was at the time little tourism development. Below is a selection of the statements with the percentage of those who 'agreed' and 'disagreed'. These percentages were calculated by adding together those who responded in the 'Strongly Agree' and 'Agree' categories to provide a total percentage of those who '*Agree*' and adding together those who responded in the 'Strongly Disagree' and 'Disagree' category to provide a total percentage of those who '*Disagree*'.

Figure 4.7: An example of a Likert scale question where results have been aggregated into the two categories: 'Agree' and 'Disagree'

Statement	Agree	Disagree
The café bar would be a good place to take friends and family	64	10
The setting up of the café/bar would encourage more tourism	42	36
The café/bar would create jobs in the area	35	46
The establishment of the café/bar would result in more drunken driving in the local area	53	33
The café/bar would disrupt family life as men will visit leaving their partners and children at home	25	49
The café/bar would be a suitable place to take children	41	36

Note, in relation to Figure 4.7, that generally the larger of the two percentages for each statement is not necessarily a very high number. In this example, only the first statement has more than 60% of the sample who 'agree' with it. Several of the statements have less than half the sample in agreement, (or disagreement), although in each case these are the bigger percentages of the two. This is not unusual in such questions and care should be taken not to overstate the point about percentages who '*agree*' or '*disagree*'. Also, remember that these are aggregate percentages. It should be possible to look again at the raw data and indicate the 'strength of feeling' in relation to each statement (in other words what percentage responded in the 'strongly agree' or the 'strongly disagree' category). As we have noted previously, there is a tendency with Likert scale questions for results to cluster around the midpoint, rather than be at the ends of the spectrum, and as can be seen in Figure 4.7, for none of the statements does the total percentage add up to 100%, this is because some respondents will have answered in the mid-point category 'neither agree nor disagree'. However, it is also important to note that when there are statements with a high percentage in the aggregated 'agree' or aggregated 'disagree' category, these will provide a good opportunity for discussion in the analysis section of your study and will need an attempt made to explain why these results occurred.

Non-standard questions

The discussion so far has concentrated on what are known as 'standard' questions. They all use words and in relation to the different types of question either a tick/cross/circle or words are required for the response. However, there are other types of question that can be used in your study. It is quite possible to use some form of illustration or visual material (it could be a photograph, a map or a graph for example) and ask respondents something in connection with this visual aid. For example, in a face-to-face questionnaire survey, you could use photographs of three different types of hotel and get respondents to rank them and then explain what factors they used to do this. A similar approach could be used with, for example, five photographs of 'landscapes that could be used for tourism experiences and respondents could indicate their preferences and explain their responses.

You could use visual material produced specifically for tourism purposes, such as tour operator brochures, or on-line tourism images as stimulus material for respondents and ask questions about the content of the images. It may be possible to direct respondents to a particular website which, for example, is a destination marketing tool and ask respondents to comment on this and to evaluate it in relation to other websites.

Another approach which has been used particularly in the marketing area but increasingly in tourism is projective techniques. Projective techniques can also use photographs, cartoons, newspapers, advertisements or the internet. A key aspect of projective techniques is that it is an indirect form of questioning that can nevertheless provoke a strong reaction. Respondents have to imagine something about the topic or about themselves. For example, one of my PhD students used projective techniques in connection with a Middle Eastern country via a face-to-face questionnaire. The student asked respondents to imagine if this Middle Eastern country was a person, what sort of person it would be. This research was conducted with samples of the same type of respondent in the Middle Eastern country and also in the UK. The results from each country's respondents were quite different and the student was able to argue convincingly that this particular technique had been the main reason for the different responses.

However, if using such non-standard techniques, it is very important to provide a clear and convincing rationale for this and explain why conventional approaches were not considered appropriate. It is also important to be aware that the results produced should be considered with great care and, where possible, should be compared with responses to any conventional questions that have been used in the research process.

Sampling, statistics and probability

In relation to most quantitative research, it is very important to decide in advance the size and nature of the group of respondents who will answer your questions and how you will get them to answer the questions. However carefully you have designed your research questions and the timing and location, it will be a pointless activity if there are major errors with how your group of respondents is selected. The nature of this selection process is very important in all social sciences research, but how you do it is related to the type of research you are conducting. In questionnaire survey research, emphasis is on representativeness and the use of a sample, and the discussion below focuses on sampling in the use of questionnaire research.

■ Sample and population

In almost every case you will be using a sample of a much larger population. But what is a sample? It is a representative group of respondents drawn from a larger population. Before indicating more about the concept of a sample it is important to state what a population is. A population can vary greatly in size. It could be the total number of people residing in a country. It could be all visitors to a heritage attraction over one year. It could be all attendees at a major sporting event on a specific day at a particular time. In each of these three examples, the actual number in the population will be vary from several million in terms of the population resident in a country, to a few thousand at the sporting event. However, the population is still relatively large in terms of conducting research

The major reason for using a representative group of the population is that when discussing your results, you will want to feel reasonably confident that what you have found out from, and about, your sample is generalizable to the overall population. In other words, your sample results are as close as possible to being identical to what the entire population would have given as responses, if you had the time and the resources to access the entire population relevant to your research.

How can you carry out research when the population you need to access is in the thousands? The simple answer, and used by almost all researchers, because of time and resource constraints, is to use a sample of the population. But the key question to answer in attempting to ensure your survey results are credible, is: 'How can you ensure your sample is representative?' Put another way, this question is: 'How can you ensure your results are a reliable reflection of what the entire population would indicate in the context of your study?' This can be difficult to answer, particularly in tourism related research, as it is frequently the case that the characteristics of the population are not fully known. For example, those visitors attending a festival who have purchased a season ticket can be asked to complete a form giving important demographic information when they purchase their ticket. However, festivals usually sell day tickets and also tickets for a part of the festival, not just for the whole festival, and both of these types of ticket will be available to be purchased 'on the day'. Is it likely that the organisers will collect demographic information on such attendees? Probably not! In addition, some of those who have purchased a season ticket, may not, for any number of reasons, actually be at the festival. So, although the organisers may have a record of the number of different types of ticket holder who attended, it is extremely unlikely that they will have details of the characteristics of all those who attended. So, they do not have sufficient demographic details on the population. Case Study 4.2, which I have used when teaching students on research methods modules, is concerned with how representative your sample is and why it is worth conducting a statistical test.

Case Study 4.2: A representative sample?

Put yourself in the position of a researcher in the West End of London who wants to gain a better understanding of what motivates tourists to go to a 'West End Show' and intends to investigate whether they are satisfied with their visit to what is generally regarded as the most important 'theatre land' in the world. Let's assume that you, as researcher, have decided a sample of 200 theatre-goers will be sufficient. You have decided you will use a questionnaire based on carefully read literature, closely related to your study, and will try to ensure a representative sample of attenders. You have decided that you will conduct the research on two separate nights in the queue outside two London West End theatres. The specific approach will be to use the technique of stopping every third person (See Mason and Beaumont-Kerridge, 2003: Raybould, 1999) in the theatre queue and handing them the questionnaire. It would appear that you, as researcher, are well prepared.

However, on the first planned night of research a severe snowstorm has led to many rail and other public transport cancellations. The audience is much smaller than usual and is mainly from within the London area. As there are relatively few attenders, the queue is small and anyway, people are not happy to complete the questionnaire 'on the spot' as it is still snowing! You are only able to achieve 62 completed questionnaires. On the second occasion, a few days later, the weather has improved, but the other theatre you have planned to use, unbeknown to you, has a special arrangement with a tour operator bringing Hong Kong Chinese visitors to London. As you reach the already lengthy queue, three buses with over 200 Chinese tourists join the other largely Chinese visitors already there. You decide to make the most of it and hand out the questionnaire as planned. Within a relatively short period of time, you have 133 completed questionnaires. However, once you have an opportunity to read through the completed questionnaires, you realise that well over 100 have been completed by Chinese visitors, who it is clear from the quality of responses, do not understand, or use very much English! How representative is your sample and hence how useful will your data be?

Case Study 4.2 indicates some of the problems in attempting to achieve an appropriate sample. However, there are several different approaches to try to ensure that your sample is as representative of the population as it can be. Ryan (1995) indicates two important aspects of this. First, the sampling unit has to be clearly identified. In much tourism and related research, the sampling unit is individuals; this can be tourists in a destination, diners in a restaurant, or users of a leisure centre. However, the sampling unit does not have to be people – it could be tourism destinations, hotels or festivals. Second, is the importance of the sampling frame and this is the list of the different units that are in the population and therefore should be in the sample (Ryan, 1995). The different units could be demographic factors, such as gender or age, or possibly geographical factors, such as specific locations, or specific types of tourists, for example 'international visitors to a heritage attraction'.

You may be tempted to use your close colleagues from university or from work or friends 'down the pub', as the major respondents in your research study. However, such people are very unlikely to be a representative sample. You may believe that you can carefully select individuals to take part in your research, but there will usually be problems with this. You may think conducting research in the street, stopping every third person who passes you is a good approach. But what about those who are at work? They will not be in the street when you are there. You may do your research in a busy urban street. But what about those who live in rural areas? You may want to find out who are regular users of a leisure centre and those who are not. You cannot assume that those who are at the leisure centre when you are there, for example, on three separate occasions, are representative of 'normal users' and those not present, when you are there, are 'non-users'!

The way to overcome these issues is by appropriate sampling of the population. Many students have heard of the term 'random sampling'. Unfortunately, large numbers seem to interpret this as meaning that they can stop anyone passing in the street (this is believed to be the 'random' part), or email whoever comes to mind (again this is understood as 'random') as possibly able to provide them with some answers. This is a misunderstanding of the term random, which has a precise meaning in statistics. I suggest this misunderstanding is also, in part, to do with certain types of sampling (which are non-random) in which specific individuals are selected deliberately or because it is easy and convenient to use them. Both random and non-random sampling are discussed below.

Random sampling

Random sampling is a statistical term, as it is based on probability theory. However, the term '*statistic*' is often used loosely, so we need to understand what a statistic is. Strictly speaking, a statistic is a numerical value, or a measure, that describes or summarises some aspect of a sample (Clarke et al., 1998). So important statistical terms include words such as *average, mean, mode* and *median*, as these are all numeric terms. When a numerical value represents a summary measure relating to a population, it is called a *parameter* (Clarke et al., 1998). Calculating the average age for a particular population would constitute a parameter. The link between a sample and a population, in relation to these terms, is that the term statistic, in a strict sense, only applies to a sample, whilst a parameter applies only to a population. In research in which you have used a random sample, what you want to be able to do is to make a generalizable prediction based on your sample that relates to the whole population. What you hope that you can write is something like: 'In my sample, the results were … therefore, it is very likely that this same result would be found in the whole population'.

Before moving on, as there has been reference above to prediction, there is a need to provide more discussion of the concepts of sample and population. Many statistics books use the term *descriptive statistics*, and the term *inferential statistic* is also commonly used.

- *Descriptive statistics* are methods used to describe or summarise data (such as the mean or median). This type of statistics often describes a group of people, or other things, using numbers, tables and charts (Clarke *et al.*, 1998). So, a group of visitors to a heritage attraction could be described according to their gender, age, nationality, and education level.
- *Inferential statistics* are methods or techniques used to make use of observed or collected data and extrapolate from this to the population as a whole. Putting this another way, inferential statistics will be based on a sample of the population but will try to predict certain characteristics and/or behaviour that would be found in the whole population. Because of the prediction aspect, inferential statistics will involve *probability theory* and will require consideration of the reliability of these predictions.

Probability theory

Probability theory is very important in relation to quantitative research involving random samples. The discussion below is not an attempt to explain probability theory in detail but outlines the theory and indicates its applicability to research in tourism and related fields.

Probability theory sounds as if it will not be particularly relevant in everyday life, but it is, in fact, very much part of it. If you use email, and regularly send messages to colleagues and friends, then you will expect to receive regular messages in return, probably on a daily basis. If you do not use email often, you will not expect to receive a large number of regular messages. So, the regular email user will expect frequent regular messages and the infrequent user will not expect regular messages. The relevance of this to probability theory, is that, depending on whether you are a regular email user, you may or may not anticipate the probability of the arrival of new messages every day. This example indicates that we often make judgements about the probability of events occurring, although, as in this case, we usually do not quantify them.

However, if we continue with this example we will introduce a numeric element – so we are now involved in quantifying. In relation to the receipt of email messages, there are two extreme positions a) you will receive messages everyday b) you will not receive messages on a daily basis. These possible outcomes can be expressed as percentages, related to different probability levels. At one extreme there is a 100% chance of receiving an email message, at the other extreme, a 0% chance of receiving a message.

Between the two extremes are a range of options. For example, you may not receive messages every day, but every two days, then we can say that there is a 50% chance (or probability) of receiving a message on any one day. If you only receive a message every third day, then the probability is 33.3%. These probabilities can also be expressed in other ways than percentages, such as using decimal values or fractions. Table 4.4 shows the relationship between percentages, ratios and decimals.

Table 4.4: The relationship between percentages, ratios and decimals (based on Clarke *et al.*, 1998)

Observed patterns	Probability Values		
Normally receives a message...	%	Ratio	Decimal
...Every day	100	1 in 1	1
...Once in two days	50	1 in 2	0.5
...Once in every three days	33.3	1 in 3	0.33
...Never	0	0	0

4

In quantitative research, we very often start from a position of anticipating that something specific will occur, and this is particularly the case when we are replicating someone else's published work or applying/testing a theory. Following on from the assumption, we can put probability measures on whether the assumption/statement is actually true or false. In relation to the example above, we have posed the question: 'What is the probability that you will receive an email message every day?' and then have given a range of probabilities from 0-100%. However, we could rephrase this question, by turning it around, so it becomes 'What is the probability that you will not receive an email every day?' In the situation when you do receive a message every day, the probability of not receiving a message is 0%. At the other extreme, the probability that you do not receive a message every day, when you are not expecting to receive one is 100%!

Why turn the statements around you may be thinking? The reason is related to the use of the scientific method in research. As we have seen in Chapter 3, this approach requires researchers to be 'sceptical' about what they will find when conducting research and instead of assuming they will find what others conducting similar research have found out, they should start from an assumption that what they believe they will find out will not necessarily happen. So, the research should be established to allow the original assumption to be tested, to find if the assumption is true or false. This means that a researcher in this position needs to create what is known as the *null hypothesis*. As we have seen in Chapter 3, a hypothesis is a testable statement. This is a statement which is related to an expectation of what you hope you will find out. It is usually the starting point of an investigation, and before conducting any statistical test, it is necessary to establish a 'null hypothesis'. The null hypothesis is usually set up to indicate that no difference will be found between different sets of data (even though a researcher may indeed anticipate there will be differences). Hence, it is a negative statement or assumption, expressed as 'we do not expect' to find differences. This explains why it is the norm to phrase probability-related hypotheses statements in this way. When testing the statement, it will be to find evidence to support the null hypothesis, or to reject it.

There is much more that could be written (and has been) about probability theory, but space in this book does not allow for more discussion of its nature. You can, if you wish, read more about the theory in your own time. However, let us turn to the practical implications of probability theory. The theory is very important in tourism and related research for these reasons:

- It assists in selecting samples;
- It is important when making inferences about the relationship; between sample statistics and population parameters;
- It assists researchers in making deductions about the differences between groups;
- It assists with predictive studies.

(Based on Clarke et al., 1998)

Taking the first point in the above list, probability theory assists in selecting different types of samples. These samples are all known as *probability* samples, or the other term for them is one we have met before in this section – random samples. In this type of sampling each person (or item, such as specific tourism destinations, or individual leisure centres) in the population has an equal chance of being selected to be part of the sample. This 'equal chance' is the key component of what makes a random sample.

Types of random samples

It is generally agreed that there is a variety of types of random samples used in tourism related research (see for example Brotherton, 2008: Long, 2007: Clark et al., 1998):

1 **Simple random sampling**. This involves random selection from a complete sampling frame such that every person (or unit, if not people) has an equal chance of being selected. However, a major issue is the difficulty of establishing the full list that comprises the sampling frame.

2 **Systematic sampling**. This is selecting, for example, every fifth person on your list.
3 **Stratified sampling**. This involves the use of key terms such as age, gender, geographical location. Long (2007) suggests that in this form of sampling it is possible to either group according to the strata (e.g. age or gender) and then systematically sample through the population or conduct simple random sampling within each stratum (e.g. age or gender), to achieve the same percentage sampled in each.
4 **Cluster sampling/multi-stage sampling**. This is when different size geographical areas are involved and/or populations are dispersed. The staged approach begins with a large area and at later stages focuses down on smaller areas. The large areas are initially selected randomly, then households and finally individuals are randomly selected.

Probability theory can help in terms of inferring parameters of a population from a sample. If a randomly drawn sample of members of a group of leisure centres in an urban area suggests that the average age of centre users is 31, given certain information about the variance in ages of the sample, it should be possible to state with some degree of certainty what the average age of all leisure centre users in that urban area is.

Differences between groups can also be ascertained and confirmed using probability theory. I use the example, later in this book, of two types of a visitor, first-time visitors and repeat visitors, to a heritage attraction. Probability theory enabled me to compare the attitudes of the first-time visitors with the repeat visitors, on a number of aspects of their visit to the attraction and confirm the differences in the results with some confidence.

Probability theory can also be used in predictive studies. Prediction is an important aspect of the tourism industry. Predicting how many people will stay at a hotel in a particular month, will be an important part of planning by the hotel manager. Predicting the number of rail passengers at different times of the day, week, month and year, will lead to the need for extra carriages or reduction in carriages, depending on the specific times, so is a very important consideration for rail managers. Such predictions are nearly always based on previous experience/past observations. So, for example rail companies and airlines use this 'past information', to create models of future demand for seats.

For your research study, you may be able to create a model using past experiences. This could be, for example, in relation to variations in demand for food and drink at different times/days/seasons in a restaurant, or demand for ski holidays in specified locations in Europe, or demand for leisure centre use during a defined period such as one month. You could build into your model some demographic factors such as gender, age, class, educational levels and income levels and take into account past behaviour of different groups. Very importantly, based on past experiences, you would also want to build into your model some probability measure on the likelihood of specific outcomes.

Sample size

We now return to the major focus of this section, sampling. Whenever a researcher starts to conduct quantitative research, one of the first and particularly important questions is: 'How big should the sample be?' For the inexperienced researcher, this question seems to loom even larger! Unfortunately, there is no immediate response that will satisfy the new researcher. The answer will almost certainly be prefixed with: 'It depends on....'. If an answer is given, the size of the sample may well be greatly above what the new researcher wishes to hear. There is certainly a strong case for having a large sample (see Long, 2007). However, it is not always clear what 'large' mean. There is a commonly held view that surveys should be 'very large' to be representative and that a small sample cannot be representative. This view may be based on attitudes to political opinion polls. In the UK and many other parts of the world, a sample of only approximately 1000 is frequently used to forecast what will happen when millions of people vote. There would seem to be many in society who are unwilling or unable to accept that such a 'small' sample could be used to predict the results of an election. However, the great majority of opinion polls are generally accurate, but it is on the relatively rare occasions that opinion polls get it wrong, that we hear about this via the media. The fact that opinion polls are usually accurate suggests that they are using a scientific approach, and, in fact, they usually employ some form of random sampling. When opinion polls do get it wrong, it is often to do with problems with the sampling frame. However, what this point about scale means for your study is, total population size is not particularly important, but the size of your sample and how you have created the sample are very important.

Figure 4.8 indicates the level of confidence you can have with samples of different sizes. This table shows, for example, that with a sample of 25 respondents, there is a 19.6% probability that your sample is not representative of the population at the 95% confidence level. Note that these figures only apply in certain circumstances, so you should certainly not aim to use the lower sample totals. The confidence level aspect is important. For example, as Long (2007) indicates, if you had a sample of 1000 in an opinion poll, at the 95% confidence level you can be confident that the true figure for support is plus or minus 3% percentage points of what the poll sample has revealed. So if one party receives 45% support in the poll sample, we can be reasonably confident that the support is actually between 42% and 48%. However this is still a fairly big range, so be aware that you should not over claim for your own results, particularly if you are using a small sample. Realistically, in most dissertation research, your sample size should be at least 50 and preferably nearer 100, if you have the time and resources to achieve this. A small sample lays you open to a major criticism that there is a relatively high probability that your sample is not representative of the population. If you are using two different groups and want to compare them, then a total of 150 to 200 is likely to be appropriate with approximately the same number (e.g. 80) in each group.

Figure 4.8: Confidence in samples (source: Long, 2007)

Sample size	Tolerance (95% probability) percentages	Tolerance (99% probability) percentages
10	30.1	40.1
25	19.6	25.8
50	13.9	18.2
100	9.8	12.9
200	6.9	9.1
400	4.9	6.5
1000	3.1	4.1

Non-random samples

This section has discussed the nature of sampling in relation to probability theory and the discussion is applicable to quantitative research. However, some other types of sample are used in quantitative research. Perhaps confusingly, these are non-random samples, but are often treated as if they are. They include quota sampling, purposive sampling and convenience sampling.

Quota sampling is easier, quicker and hence less expensive to conduct than random sampling. However, it involves the identification of certain segments of the population (e.g. women aged 25-34 and/or men aged 45-54). Interviewers in the street often use this approach, having first asked a screening question to identify that the person they have stopped fits the appropriate category. However, this is not random sampling, as the interviewer can stop and use anyone who fits the quota (i.e. there is not 'an equal chance' of all potential respondents from a population being sampled).

Purposive sampling is when specific individuals are targeted because of who they are, what they know and how they are likely to respond to the questions. So, this could be international tourists in an airport departure lounge waiting to take an international flight, or waiters from a four-star hotel chain. Purposive sampling is often used by students (although it is not always referred to as such) because it involves respondents who are viewed as particularly relevant to the research, because of for example, their knowledge of a particular topic, rather than using random sampling.

Convenience sampling means very much what it says. It uses just about anybody who is available at the time and in the place at the time. By definition this form of sampling is not likely to yield results that are particularly meaningful. However, it seems to be used by students and other students are frequently the 'convenient' sample. If you are to use this approach, you need a very convincing rationale and need to remember that it is non-random, so probability theory does not apply, and you cannot call your sample representative. It may be suitable to use convenience sampling in a pre-test, but you should think very hard before using this as your main sampling approach

Finally in this section, samples are also used in qualitative research, although some researchers do not like to use the term sample as they consider it as only applicable to numeric-based, random sampling. However, these are samples which are not affected by probability theory and are known as non-random samples. For example, purposive sampling is used in qualitative research and a number of other non-random sampling techniques are also employed in qualitative research. These are discussed in Chapter 5.

Writing this section in your study

In your study, the discussion of your overall research methodology and specific techniques would be best located in what is likely to be Chapter 3, which would probably be titled: 'Methodology'.

If we assume that you have a Chapter 3 in your study entitled 'Methodology', then the first part of this, as indicated in Chapter 3 of this book, would discuss the overall research philosophy you had used in your research, before moving on to discuss the specific techniques you have used.

If we assume that you have used a questionnaire in your primary research then Figure 4.9, which is a summary of preparing a questionnaire survey, should be helpful in not just the design and use of the survey, but should act as a guide to writing this section of your Methodology chapter.

Figure 4.9: A summary of questionnaire preparation and use.

- *Ensure you inform your respondents.* You will probably need to provide some form of covering letter to explain who you are and the aims and objectives of your research, as well as informing respondents how you will deal with their answers to your questions. Specific questions may require specific instructions.
- *Make your questions as clear as you can.* If it is an important topic, you may need more than one question, but do not impose irrelevant questions on your respondents as this is a waste of time for them. If your questions are not clear, your results are unlikely to be reliable.
- *Avoid certain types of question.* These include hypothetical questions ('what if…?'), leading questions and ambiguous questions. Also check on the advantages and disadvantages of the actual questions you have used, e.g. whether closed-ended or open-ended.
- *Do not use two questions in one.* Do not ask questions such as 'Have you been to this theme park before and did you enjoy your visit?' (see also Case Study 4.1 above).
- *Do not make respondents take up extreme positions.* Do not ask use a phrase such as: 'International hotel chains never pay adequate wages' before asking the question: 'Do they? The respondent will find this difficult to answer, particularly if they actually believe there are occasions when this is true, but it is not always the case.
- *Do not ask sensitive questions* which are likely to lead to emotional responses. Direct questions about age, educational attainment or income may annoy respondents. There are other topics that respondents may regard as sensitive. Be aware of possible sensitive topics and consider whether you need to ask such questions! If you do ask this type of question, your respondents may not answer accurately.
- *Do not use jargon/terminology that your respondents may not understand.* A question such as 'Cohen used a four-fold classification of tourists: mass tourist; individual mass tourist; explorer; drifter. Using Cohen's terms, how would you classify yourself?' This question cannot be answered unless the respondents understand what the categories actually mean!

- *Do not ask questions that require a detailed memory of events*. Do not ask this type of question: 'How many times per year did you take a holiday: Last year, 5 years ago, 10 years ago, 15 years ago, 20 years ago?' Many respondents will not have an accurate memory of 10 years ago, let alone 15 or 20 years ago!
- *Do not ask vague questions* such as 'Would it be likely that you might visit a theme park next year?'
- *Do not ask respondents to make detailed calculations before they can answer*. Consider the question: 'What proportion of your total holiday budget are you likely to spend on having meals out with your family?' This is about something that has yet to happen and the respondent may never have considered such a calculation themselves. Therefore it is very likely that their answer will not be accurate.
- *Separate Categories*: Do not use overlapping categories. For example, do not use age categories: 30-40, 40-50 50- 60. You should have e.g. 30-39, 40-49, 50-59.
- *Structure of the questionnaire*. There should be a progression from 'easy to answer' questions near the beginning, followed by more difficult ones in the middle and some easier ones towards the end.

With reference to Figure 4.9, you can indicate in your writing of this section of your 'Methodology' chapter how you acted on the advice provided here. For example, if relevant to your particular questionnaire, how you avoided the use of sensitive questions and also ensured that categories did not overlap.

What you write in this section of this 'Methodology' chapter will be clearly linked to the specific techniques you have used. When discussing the technique (s) you used, one of the key points is that you should not just describe the use of the technique, but you will need to provide the rationale for the selection and use of the technique. What you are doing here is making an argument for the use of the particular technique that you used. You will also be providing an indication of why you did not use another technique.

When discussing the specific questions, you have used on your questionnaire, once again you will need to provide a rationale for this. This will involve discussion of the specific types of questions you have used and the sequencing of these questions.

In the discussion of the rationale for the choice of the technique, the nature of your questions and the sequencing of these you will need to provide literature references to support your statements. As questionnaires are used throughout the social sciences, there is a large amount that has been written about, for example different types of questions, and the order in which these can be presented. You need to draw on as much of this literature as you can, to provide the support for your selection of technique(s).

In the writing up of how you selected your technique(s), you should provide, in addition to what has been discussed, confirmation that you have considered, in advance, how you will analyse your specific questions. If it is your intention to use a computer package, then make sure beforehand that your question can be analysed using the computer package and indicate that you have also considered this prior to using your specific questions.

Ethical issues will almost certainly have to be discussed in this section, whatever technique(s) you have used in your primary research. Key areas you will have to refer to are confidentiality and anonymity. You may also have to indicate how you will store completed questionnaires, if used, to ensure security of data. There may be other ethical areas, perhaps in relation to your own safety and security that you will need to consider. Also, for whatever type of quantitative technique(s) you used, explaining how you obtained 'informed consent' from participants will be important.

Student activities

1 Under what circumstances would you use closed-ended questions on a questionnaire survey?

2 Under what circumstances would you use open-ended questions on a questionnaire survey?

3 Why are attitudes an important area of research? What problems can occur when you are attempting to measure attitudes?

4 What are the advantages and disadvantages of Likert scales as a technique for measuring attitudes?

5 In relation to Case Study 4.1 study the student's first draft of the questionnaire and the supervisor's comments and produce a second draft of the questionnaire.

6 What are the advantages and disadvantages of aggregating Likert scale responses into the two categories: 'Agree' and 'Disagree'?

7 What is the difference between a 'sample' and a 'population'?

8 What is the meaning of the term 'random sample'?

9 What differences are there between descriptive statistics and inferential statistics?

10 In Case Study 4.2 what are the problems with the sample that has been obtained?

11 What are the links between probability theory and prediction?

12 What is the 'null hypothesis' and why is it important in quantitative research?

13 What factors will affect the size of the sample you use in your in your primary research?

14 Referring back to Chapter 3, indicate how ontology and epistemology are linked to the quantitative techniques presented in this Chapter.

5 Qualitative Research Techniques and Conduct

Introduction

Until fairly recently, many students were concerned about using qualitative research. This was usually due to of a lack of confidence, or a lack of experience of using it, or both of these factors. The concern may also have been a result of a tutor not feeling confident in the student's ability to use this approach. This, in turn, may have resulted from the tutor's concern about their own ability in using a qualitative approach.

Another reason is that students may feel that qualitative research is 'inferior' to quantitative research. Such students (probably based on their tutor's reaction, other students' comments, or on their own reading) will hold the view that such research is 'soft' in comparison to quantitative research. The order of chapters in this book, and most textbooks on research methods, also reflect the conventional wisdom on qualitative research, that it is 'secondary to', and perhaps supportive of, quantitative research.

However, this chapter is presented to suggest that qualitative research should not be viewed in this way – it is an approach that is useful in a variety of contexts and should not be viewed as secondary (i.e. at a lower level than) to quantitative research. Rather, it should be viewed as an alternative way of gathering information and, in fact, students should be aware that many researchers view it, not as secondary to, but at least an equal footing with, or indeed better than quantitative approaches.

Qualitative research

As earlier sections of this book have indicated, using a qualitative methodological approach is closely related to the way we individually know about and view the world. Hence, ontology and epistemology are very significant in guiding qualitative methodology, in the same way that these factors also guide the use of a quantitative approach. If your research is underpinned by the idea that all individuals view the world differently, but that these views each have value and by comparing these views you are gaining understanding about an issue or topic, then a qualitative approach will be appropriate.

Also, if there is little apparently known about your specific chosen topic (i.e. there is a lack of literature on the subject), there is unlikely to be much in the way of theory to apply to your research topic. Your research is therefore likely to be, as we have discussed previously, inductive rather than deductive. This means you do not start your research by applying, or testing a theory, or a number of theories in the field, but instead collect data. What you have collected in the field can then be used by you to build up theory and then this can be compared with what may have actually been published on the topic, or a closely related research topic.

Research which is concerned primarily with the individual views or attitudes of respondents, stated in their own words, is suitable for a qualitative approach. If you are intending to gather responses which will be expressed, primarily, in words and these are the words chosen by your respondents, rather than by you as researcher, then a qualitative approach is appropriate. A very common way that students gain quite lengthy word-based answers, expressed by respondents in their own words, is via interviews. Therefore, interviews are generally regarded as a qualitative research technique.

For some students who are not that familiar with the philosophy of research, the division between qualitative and quantitative research may seem artificial any way. It may also be that some research techniques usually regarded as qualitative are 'common sense' to you. For example, using an interview is usually regarded as employing a qualitative technique. Interviews are also commonly used in a variety of media such as TV or radio, so you are likely to be familiar with interviews, even if you have not used this approach. However, you may have used an interview approach before, at a lower level that your current research, in an attempt to gain a detailed insight into a respondent's experience of, understanding of, or views on a topic. At the time you may not have been aware that you were using what is commonly regarded as a qualitative technique – this was just the advice of your tutor or teacher, or you had read an article where a researcher was involved in a similar study to yours, so followed their approach.

As noted previously, an important difference between qualitative and quantitative research, is that one tends to be deductive while the other tends to be inductive. What this means is that in quantitative research the person conducting the research is frequently making use of a well-known theory and probably applying it in a new situation, such as with different respondents or a different location. For example, the work of Agarwhal (1997) applies the theory of Butler (1980) on the Tourism Area Life Cycle (which has been discussed previously) to a particular location, Torbay in the UK. Agarwhal was able to apply Butler's theory and compare her results with what Butler indicates in his theory. This type of quantitative research may also enable the researcher to predict what might happen in the context of their research, based on what a particular theory indicates. In such circumstances, a researcher can make use of hypotheses – as we have seen previously these are statements that can be tested.

However, most qualitative research is not like this. Qualitative research, involving perhaps interviews or focus groups, is being used because it is not known or fully understood what will happen in the specific context. Usually this means that little or no research has been conducted in relation to the topic. Or it could be a researcher is examining an old topic, but from a new, unusual perspective. In this case, the researcher does not want to prejudice the research process by asking questions in the way that they have been asked before. If the researcher uses the same questions and the same technique, it is very likely that the results will be similar to those of others who have used these questions and this technique. Therefore, in a qualitative approach, questions are posed which may then help to contribute to the development of new theory, rather than being based on existing theory.

Although the two research philosophical approaches are often thought of as different and separate, a good deal of research in tourism involves both qualitative and quantitative approaches. It is often the case, in terms of a new research theme or the re-examining of an old topic from a new angle, that a researcher starts from a position of being as open-minded as possible. Therefore, it is likely to require a qualitative approach in this part of the research. However, a researcher may want to follow up relatively open-ended research, which has involved only a few respondents, who were asked relatively few questions but required to provide in-depth, detailed responses, with an approach in which more respondents are involved. Here respondents will each be asked more questions than in the initial qualitative phase of the research, but responses are more likely to be short answers and/or closed-ended. This is, in fact, an approach taken by some PhD students in particular and is known as a mixed methods approach – this approach is discussed in more detail in the next chapter. It has the advantage of allowing comparison between the techniques at different stages of the research and hence the results. It is likely that it will also allow the disadvantages of one technique to be offset or cancelled out by the use of other techniques. In this way, results could and should be more reliable and the researcher can feel more confident that they are not chance findings, or that there was something unusual in the process or the actual results.

As indicated above, qualitative research is best conducted using techniques which are unlike those used to collect quantitative data. The words used by respondents, as opposed to their 'ticks' and 'circles' on a questionnaire, are the major 'data' gathered in qualitative research. A very important way to gather this data is via interviews. Focus groups are to a very large extent group interviews so are another way to gather qualitative data. Another technique used in qualitative research is observation. The techniques of observation, interviews and focus groups are discussed below.

Observation

Students are generally aware that they need to ask questions when conducting any form of research. However, when conducting primary research in the field, this often translates into the feeling that they must have respondents in front of them and ask these people questions. This is frequently a result of concern to get the most from those people involved, based on an awareness that time with respondents will be limited. What students may not realise is that it is not always necessary to ask respondents questions directly. It is possible to conduct successful research while merely observing what those being researched are actually doing.

Observation is a process each one of us is involved in on a day-to-day basis. We see (observe) and speak to our friends, look at what they are wearing, watch what they do and listen to what they say. Almost everything we do requires some form of observation. Therefore, because observation is a regular activity, it should not be that difficult to transfer this process to a research context.

If you drive a car, then you may well have heard of the acronym COAST. The initial letters stand for Concentrate, Observe, Anticipate, Space and Time. The O in COAST is clearly important when driving, but also the other letters in the COAST acronym can be seen as important in the research process of observation, as you, as researcher, will need to concentrate and anticipate, while observing, and whatever you are observing exists in space and time.

Observation, however, is not just about using one's eyes but also ears and could involve a sense of smell and even touch, depending on what is being observed and the context for this.

In research terms, there are two major types of observation: participant and non-participant (Veal, 2011).

- **Participant observation** is usually 'declared' (overt). This means that those being researched about know what the researcher is doing.
- **Non-participant observation** is usually conducted as an outsider and can be 'non-declared' (covert), meaning that those being researched do not know that the researcher is studying them. However, the ethics of conducting such research should always be considered carefully.

The structure of observations

Observation can be conducted in a very structured way or be relatively unstructured. In a specific setting, it may be reasonably easy to anticipate or even ascertain the possible actions that could take place, and these may be limited in number, so it may be possible to create an observational check list in advance of the research encounter. In other situations, for example when little is currently known about what could occur in a particular situation, it may well be better to keep an open mind and not have decided on specific activities to observe and record in advance. This approach has the advantage of allowing the observation of certain activities and events that would go unobserved if specific activities were the main focus of what was being studied. Such occurrences may, in fact, be important in the research. However, it has disadvantages compared to observation with a specific focus, in that the attention of the researcher wanders and what is important is not observed at all or is not clearly understood. In practice it may be a good idea to begin observing an activity with a fairly open mind and then after observing it over a period of time, or on several occasions, a more structured observational schedule can be created.

The initial unstructured phase of the observation could be in the form of a pilot study, with the objective of developing and trialling specific questions giving direction to the observation – in Chapter 3 there was detailed discussion of the nature of pilot studies. Alternatively, the unstructured observation could be part of research which is modified and adjusted over time according to what occurs during observation and the results that are obtained. In both of these contexts, recording what occurs and providing a rationale for any changes or modifications are very important. Remember both your supervisor and the marker of your dissertation will be very interested in not just how you conducted your research, but why you did it in the way that you did!

Students are often concerned that the way in which they conducted parts of their research may appear 'messy'. But you should remember that all research, particularly qualitative research can be considered as 'messy', although journal articles may not always reveal this.

Observation is very flexible as a research technique in that it can be used in a variety of settings from the very formal, (such as an organisation's regular meeting in which minutes are taken, or a pre-arranged sporting event), to very informal gatherings of people such as on a summer's day with groups of visitors, for example, on a beach, or by a lakeside or by a swimming pool. This flexibility means that observation can be used, for example, to count the number of users of a piece of gym equipment, but perhaps less formally to study participants' lifestyle at a particular event or on a specific occasion.

Observation does not have to be a 'one-off study' of, let's say, a group of spectators at a football match. It can be conducted over a period of time, with day-by-day observations. In fact, it can be used over months or even years. The biggest single observation in the UK was known as 'Mass Observation' (see Long, 2007). It took place from the 1930s to 1950s with a focus on observing largely working-class life. It was concerned primarily with the working lives of thousands of British people, but also with leisure activities. Much of the research was centred on 'pub culture' – what occurred in British public houses involving those being studied. The observational process required an observer (and there was a very large team of observers in Mass Observation) to spend a lot of time regularly with one group of people.

In the case of Mass Observation, the observer also engaged in conversation with those being studied and this became part of the research process, as well as the observation itself, specifically to discover the participants' interpretations of the events that the researcher had observed. It may be appropriate for you to use observation, plus some form of questioning of those you observe, in your primary research for your dissertation. For example, it could be very interesting to note that what some respondents tell you, is not the same as what they actually do when you have observed them! Hence combining these two approaches is intended to lead to a more comprehensive understanding.

The role of the observer

The role of the observer can vary greatly, particularly in terms of the researcher's involvement, or lack of involvement, with those being observed. The observer can attempt to not participate at all in what is being observed, but this will be difficult, as even the act of observation can alter the behaviour of participants. However, if the observation process is covert, bearing in mind the ethical issues of this, it may be possible to achieve this.

Often the observer needs to be a participant, such as being in a crowd at a sporting event. In this case, the observer does not need to remain separate from what is being observed. Another type of observer could be very actively involved as a participant, such as joining a football team supporters' club and attending all matches with a regular group of fans over a period of several games (see Parks, 2003). This form of observation should give much deeper and richer insights into the behaviour of participants, but clearly will require a much greater investment of time and probably effort.

What are you looking for?

Whatever observational approach you use, the key question is: 'What should you be looking for when you observe?' Some of the possible questions you can use when observing are presented below.

- Who are the actors? In other words, who is involved?
- What are the actors doing?
- What effects does this have (e.g. on other actors, the context in which events occur)?
- How do the actors interact with each other?
- Are there any particular rules of behaviour?
- What do the people observed/the actors look like (e.g. dress, facial expressions)?
- How do they speak?
- What non-verbal gestures/body language can you observe?

When involved in the observation process over a period of time, you will need to decide on precisely what to concentrate on and for how long. There may be a tension between taking a closer look for a short time and taking a broad look for a longer time. However, this process may assist with attempts to focus your observation questions and give you more structure.

If your observation is overt (those you are observing know what you are doing), you must consider your impact as observer on the participants. Think about whether your presence affects the way your participants behave. You should also be aware of your responsibility to those being observed in terms of letting them know precisely what you are doing.

If, however, you are involved in covert observation, you should consider implications for the quality of your research if your position is compromised. In other words, you need to consider, if your participants discover you are an observer, what effect this will have on your research. Also, as a covert researcher, you should also think whether there could be a possible danger to your safety if your real identity is revealed. This may seem very unlikely but consider what could happen if you have been closely observing a group of football fans, who believe you are also a fan of 'their team', and you have been studying how they behave when their team wins or loses. What would happen if it became known to the fans that you are not really a fan of their team, the team has just lost and they decide that you are in fact a fan of the opposition, who have just beaten their team?!

Advantages and disadvantages

Observation has certain advantages, in particular that it is of a very personal nature, in that the researcher and research instrument are one and the same thing. This limits the chances of things going wrong. In addition, the researcher may be in a position to conduct the observation on several occasions, which should help with the reliability of results. In this way observation may be better than a questionnaire survey which is likely to be little more than a brief 'snapshot' in time and space.

Nevertheless, this apparent advantage is potentially a disadvantage as observation can be criticised as it tends to be very subjective – it is largely the views of one researcher recording what happens. Hence, the reliability of the observation can be questioned. Observation often involves a small 'sample' of a population, so again the reliability of the results can be questioned. As the process is relatively subjective, there is also an issue with the way in which results should be analysed. These criticisms should be borne in mind when considering using observation.

A way to offset these issues is to conduct a pilot observation and make modifications to the observational process. In addition, the way in which results will be analysed needs to be given serious thought before the main observational process. If the analysis process is not considered in advance, a large amount of data may be collected, but it may be of little use.

Recording data

However, there will still be an issue with how you record your observations. You can take notes, manually, or use some form of audio recording device such as a tape recorder, Dictaphone, MP3 player or mobile phone. Alternately, you could use a stills camera or a video camera. Whatever method you select, you should be aware of the impact it could have on those being observed. If you are conducting covert observation, you will need to consider, when using your selected recording technique, whether it is possible to remain a covert observer.

It may be possible to conduct such observational work without respondents knowing that they are the subject of research. This can have advantages – respondents usually act more naturally when they do not know they are being observed for research purposes. But it may also be appropriate to inform participants that they are being observed for research purposes, although their informed consent would be required in this approach. Therefore, if using a covert approach there is an important ethical dimension which will need to be addressed in appropriate sections in the dissertation.

As indicated briefly above, observation does not have to be used as the only research technique. It can be used in conjunction with other approaches. Hence, observation can be used to supplement direct questioning of respondents. In this way the results of observation can be compared with those from another approach, such as an interview or a questionnaire. By comparing, similarities in results of the different approaches may be seen and this is likely to make results more reliable and differences can also be recorded. For example, as indicated above in relation to Mass Observation, it could be very interesting to note what some respondents tell you, in relation to what they actually do when you have observed them. Hence combining these two approaches is likely to lead to a more comprehensive understanding. Figure 5.1 below provides a summary of the different approaches an observer can use.

Figure 5.1: Different forms of observation

- The complete observer – This is an approach in which the observer tries not to participate at all. This is similar to a police interview in a police cell, when other police officers look through a one-way mirror at an interview, with at least one police officer and someone accused of a crime. However it is very difficult to achieve non-participant observation in reality, as the observer will frequently affect who is being observed.

- The observer as participant – This involves a formal observation (e.g. when an observer attends a sporting event as a spectator)
- Participant as observer – An example of this would be when the researcher 'hangs out' with a group of football fans.
- The complete participant – In this case, the researcher 'lives the life' of those being observed. For example, the observer becomes a football fan and attends all the matches, visits the supporters clubs, and goes to public houses and bars with the fans over a lengthy period such as a complete season of games.

Case Study 5.1 is based on the work of students who were studying at a university in London and was part of the course on research methods. I used this exercise as preparation for their postgraduate dissertation. It indicates different types of observational processes and subjects for observation.

Case Study 5.1: Observation at Stonehenge

Background

Stonehenge and Avebury are major heritage attractions in SW England. They are both World Heritage Sites. Stonehenge has approximately one million visitors annually, with a majority of visitors being international, while Avebury has less than 300,000 visitors, most of whom are domestic. Stonehenge has relatively high entrance charges, while Avebury has no entry charge. Car parking is one of several important management issues at each site. The observational fieldwork was intended to look at the issues confronting each of the two sites and the related management approaches

Stonehenge and Avebury Observation Schedule

1 Upon arrival in the car park, but before entry

For a 15 minute period, record the number of different types of vehicle in the car park.

For a 15 minute period, record the number of the following visitor group types. School/university groups, family groups, coach trip arrivals, couples, others.

2 For a 15 minute period, record where visitors go first on arrival.

(Stonehenge only) How long is the queuing time?

Once you have entered the site

(Stonehenge only) What percentage of visitors use the audio guide?

(Stonehenge only) How do visitors make use of the audio guide? (e.g. continually/occasionally/very little)

What do visitors do once on the site? (Take photos? Just walk? Listen to the audio guide? Talk to others?)

How long do visitors stay at the site?

3 In a 15 minute period, how many visitors go to the souvenir/gift shop?

In a 15 minute period, how many visitors use the cafe?

Listening to visitors

4 Where are visitors from? What evidence have you used?

5 What comments do you hear from visitors on the following (Do not ask visitors, but politely listen to what they say):

- Quality of facilities/toilets?
- Value for money
- Their overall experience

Your views

6 What is your view on the number of visitors to the site (Is it overcrowded? empty?)

7 What is your view of the following:

- Quality of facilities
- Value for money
- The overall visitor experience

8 Any other comments?

Interviews

Interviews may seem to be the most 'natural' or 'normal' form of research that you could use in your research project. Talking on a one-to-one basis is something almost all of us do every day. This frequently involves asking questions of friends and family members. You are no doubt aware of interviews being conducted in a variety of media, including TV and radio. Newspapers and magazines make use of the interview format with celebrities and others, such as experts, to gain their opinion on a variety of topics.

However, you should be aware that the type of media interviews you know about will not be of the same format as using interviews in your dissertation research. Although there is a large range of interview styles and formats, the type you can use in your study will be formalised and should have set frameworks and structures, although there is some flexibility within these frameworks.

In terms of research, interviews can be regarded as one more possible technique that you could use. However, interviews are generally used in specific circumstances and are seen as having advantages over other techniques. The discussion of questionnaire surveys in Chapter 3 indicated, in addition to several advantages, that there are a number of disadvantages. One key problem with the format of questionnaires is that the questions used do not allow in-depth responses. They generally do not provide an opportunity for respondents to give a detailed explanation of an answer to a question. Also, they do not usually allow further exploration of a topic. As an interview involves a 'real-time interactive' process, the disadvantages of a questionnaire indicated immediately above can be overcome.

However, interviews are not just used as an alternative to other possible approaches; they are frequently perceived as being the best way to get useful research material. In fact, the first inclination of a very large number of social researchers will be to conduct interviews. A major difference between interviews and questionnaires is that the researcher using an interview wants to discover how the respondent views the world and how they express this in their own words, while the questionnaire user is much more concerned with respondents' views on what the researcher considers as important (Long, 2007). This can be seen in the way that the researcher using a questionnaire, designs and prepares questions which usually have short closed-ended response categories, while the researcher using an interview technique, although using specific questions, will allow respondents to go into detail, provide explanation and add extra comments in their own words.

The interview 'sample'

Interviews are used mainly to collect 'rich' or 'thick' data. As this is usually in the form of words that are those of the respondent, or interviewee, this data is primarily qualitative material. As an interview can take a relatively long time – from 20 minutes to just over an hour is common (see Veal, 2011) – only a small number of respondents are usually involved. Often, this may be between 10 and 20 respondents or, depending on circumstances, less than 10 and very occasionally only one interviewee (see Cohen and Manion, 1995). What quantitative researchers might claim to be a relatively small 'sample', is made up for in the very rich data that can be gathered from e.g. 20 respondents who each provide detailed answers during a one-hour interview, meaning there is 20 hours of material.

Although in the above paragraph, there is reference to a sample in relation to qualitative research, some qualitative researchers are unhappy with the use of the term sample. The main reason for this is, as we have seen in Chapter 4, the term sample has a specific meaning in research and relates to numeric factors. The qualitative researchers who are unhappy with the term 'sample' argue that they are not interested in gaining a set number of specific respondents who are representative of a total population (in other words a random sample), hence the use of the term is not appropriate for their research.

However, qualitative researchers do use groups of respondents and often they are carefully selected. This has led some commentators (see for example Long, 2007) to argue that qualitative researchers do in fact use samples – it is just that these samples are not usually random and are not necessarily called samples by the researchers. A good number of qualitative researchers will confirm that they use non-random sampling which includes purposive and convenience sampling. In Chapter 4 there is discussion of both these types of samples. Briefly stated, purposive sampling involves carefully selected respondents who fulfil a particular role in relation to the research aims or objectives. So, a purposive sample could comprise for example 'International tourists visiting a World Heritage Site in the UK' or 'Females aged 21 to 40 attending a leisure centre in City X for a weekly fitness class'. Convenience sampling means very much what it says, in that the sample is made up of those who in a particular context at a particular time. In other words, they are at a specific location or on-line at a similar time and are able to take part in the research when required. It should be clear from the discussion in Chapter 4 that both types of sample, be they purposive or convenience, will not be a representative sample. However, as Patton (1990) argues in relation to purposive sampling (and this can be the case, but is not always, for convenience sampling) those who have been selected are not there because they are representative of the population, but because they are information-rich in relation to the research topic.

There is another approach which is used by large numbers of qualitative researchers and is often employed by students. This is 'snowball' sampling (see for example Ryan, 1995). This is when your interviewee provides the contact details of other potential respondents who they believe will be able to provide useful answers. Each successive interviewee may provide several potential respondents and you can then select your interview respondents from those suggested. The reason for you asking your respondent to supply details of other potential interviewees, is that they will be giving you details of those they believe are 'information' rich. Snowball sampling is therefore another example of purposive sampling.

An important factor in relation to the nature of interviewing, and why the number of respondents is usually relatively small, is that analysing interviews is a lengthy process. The usual way to do this, assuming the interview has been recorded electronically, is to transcribe the material (type it up) before analysing it. It is commonly accepted that one hour of electronically recorded interview will take three hours to transcribe and clearly more time to analyse (see Cohen and Manion, 1995). It is not difficult therefore to understand why 20 interviews may take up to 100 hours in total time to conduct, transcribe and analyse.

The nature of interviewing

It is important to be aware that interviews are not an easy option, particularly in relation to other research techniques. Students are likely to believe that they will need confidence to be with and ask questions of people who they have never met before. Although this is important, one of the skills needed in interview creation, is the same as that required in the design of questions for use with a questionnaire. This skill is to create specific questions for your interviews, but also an overall structure to your questioning, that has a logic to it that the respondents can understand and follow, but also meets your aims and objectives. Another major skill required, which is not one needed for questionnaire design and use, is the ability to listen to respondents. So, the talking, via questioning, is not as important as the listening to the answers to your questions.

It is not just listening, as you might do to family members or friends, in a passive way, but active listening that is required. The researcher needs to ensure that the respondent quickly relaxes and gives detailed, useful answers. This may require body language cues and verbal prompts to get the respondent to provide a more detailed response than they have initially provided. The prompting may also be an attempt to get the respondent to provide an explanation for their answer. You need to recognise the meaning of what you are listening to, ensure you can lead the discussion in the direction you want it to go and ask relevant follow-up questions. This means that you may need a well-structured interview schedule, although some research interviewers may have a fairly unstructured, open-ended approach. Even in this less structured approach, it is still highly probable that the interviewer wants to get comments from the interviewee on certain topics, which the interviewer has set in advance. The 'unstructured' interview schedule in this case, largely means that the order of the questions can be varied, but the question topics need to be covered during the interview.

It is generally accepted that there is a continuum in which questionnaire-like, open-ended questions arranged in a set sequence are found at one end and, at the other end of the continuum, are interviews with almost no set questions which are intended to replicate as closely as possible an informal chat between friends (see Long, 2007). However, you should be aware that although 'the friendly chat with mates' seems a suitable approach to ensure relaxed respondents, it will be very easy to wander from the focus of your research and then it is likely the interviewee will think: 'Why I am here?' and, at the end of the interview, you may have little of use to transcribe and analyse.

Most researchers that use interviews will have some form of a checklist of major topics, even if they do not have precise questions. However, it is probably best, as an inexperienced researcher, that you should actually have an interview schedule with questions typed, on your electronic device, or written out in full. An important reason for this is that if the interview takes an unexpected turn, and goes in a direction you had not planned, you will be able to look at your set of questions and bring the interview back to your original topics. However, you also need to be flexible and allow your respondents to not only answer in their own words but do so in the order they wish to. It is not a good idea to stop someone who is answering what you have as, for example, Question 3 on your interview schedule, with what you would consider as answer to Question 5, by saying 'Hey, wait a minute we have not got to that question yet!'. Under these circumstances, you need to allow the respondent to continue. When they have stopped responding to the specific question, it is then that you should glance at your schedule of questions and see how you should proceed. Being able to allow respondents to answer in their own way but also control the direction of the interview, is a skill that may require practice, as in may not be as straightforward as it initially appears.

Interview questions

In terms of the nature of questions for your interviews, the advice that has been previously provided for questions in a questionnaire survey generally applies to interviews. The way that you create the questions can come from a variety of approaches. So, you can start by looking at key pieces of academic literature, or if your theme is topical, perhaps newspapers, magazines TV, radio or the Internet will be places to go for inspiration. You may also find it a good idea to initially brainstorm topics and themes for your questions and then arrange the question topics in some kind of sequence.

As with advice on survey questions, you should avoid questions that are leading, ambiguous, long and convoluted, and ones which respondents may regard as too sensitive. Unlike a questionnaire, there is little point in asking closed-ended questions – unless you intend to follow up with a supplemental question. For example, if you asked: 'Do weather conditions at a destination affect your holiday choice?' The closed-ended response categories are likely to be 'Yes' or 'No'. In an interview, you would probably want to follow this up with: 'Please explain your answer'. However, it would be better to ask a question such as: 'How does the weather affect your choice of a holiday destination?' This is an open-ended question which allows respondents to give an indication of the extent to which weather affects their choice. They can give an answer such as: 'Not at all' or, 'To a great degree'. Alternatively, they could provide a comment which is anywhere between these two answers. They may also use examples to explain their response. For example, a response to the question could be: 'I am not bothered much about the weather when I spend a few days in London around Christmas, but when I go to Majorca in June, I would be unhappy if it was not hot, dry and sunny'. If they did not give an example, you could prompt them to do this, to help explain their answer.

In terms of sequencing of questions, it is a good idea to start with relatively straightforward factual questions, if you are asking such questions. You may have some demographic questions – assuming that you have not, prior to the interviews, already selected your 'sample' on the basis of demographic information. If you require demographic information, then questions concerned with this area could be asked at the beginning of the interview (although these could instead be left until the very end of the interview, as they should be easy for respondents to answer). The more complex detailed questions, requiring some time for respondents to answer, should be placed in the middle of the interview schedule, when respondents have 'warmed up'. Towards the end, more straightforward questions, such as factual ones, should be provided to bring the process to an end, relatively quickly. At the very end of the interview, you can also ask respondents if they want to ask you anything or wish to add further comments on any aspect of the interview.

Recording the interview

Unlike with most questionnaires, the respondent in an interview does not have a form to complete. The job of recording an interview will be your responsibility as researcher. It is now generally accepted that some form of electronic recording device is particularly useful in interviews (Long, 2007; Brotherton 2008; Veal, 2011). It is very difficult to keep an interview going while writing down what a respondent says, and the respondent does not want to hear you keep saying: 'Could you repeat that please?' If this occurs, it will also stop the flow of the interview, so your data may be of a lower quality. However, some people may object to having their comments recorded, particularly if you are dealing with a sensitive or controversial topic, as they may believe you will replay what they have said to someone in authority, such as their manager. Yet other respondents may see the recording as an opportunity to act as if they were a media celebrity and hence not necessarily provide truthful responses. If, however, the recorder is small and is placed discreetly away from the direct gaze of the respondent, interviewees will probably answer fairly naturally.

I also suggest that you have with you a notepad or paper and clipboard with you to write on during the interview. This gives the impression that you are a professional, but also that what the respondent has to say is important (i.e. important enough to be written down). You could use your notepad to indicate body language of the respondent, such as visual expressions in relation to questions as this will not be recorded elsewhere, or you could write down key points they have made or could write a reminder to yourself of a topic that you wish to return to in the interview. For whatever purpose you use your notepad, in my experience, it provokes a better response and higher quality answers from your interviewee, than just using an electronic recording device.

As indicated above, a key skill in interviewing is being able to flexibly respond to the interviewee's comments and also, when they fall silent, to get the interview moving again. This is not a skill that is usually required when talking to your friends and family. When you talk with them, you will naturally allow silences for, at least, a short period of time. This will not be appropriate in a formal research interview. You are the person who is responsible for keeping the 'conversation' going in the interview context. You will need to insert your questions into this conversation but do so in such a way that the discussion continues to flow and keeps the interest of the interviewee. This means that the interview will require preparation. You will need to have a very good idea of the questions you want to ask and, when there is a suitable opportunity, put the question to the interviewee. If the conversation generally flows in the interview, it is likely that the interviewee is interested in what is happening and will provide you with useful answers. If the conversation continually breaks down, and you have frequent problems reviving it, there is a strong likelihood your interviewee will lose interest and consequently your data may be poor.

A number of researchers (see Patton 2002; Whyte 1982; 1984) have suggested ways to maintain the interest of your respondents and keep the conversation flowing. Figure 5. 2 shows Long's (2007) summary of these.

Figure 5.2: Long's summary of interviewer responses and activities

Response	Activity
Non-verbal response	e.g. 'Uh-huh' indicating interviewer is listening and is interested
Verbal encouragement	e.g. 'That's interesting' encouraging interviewee to carry on
Reflection	Interviewer repeats last statement of interviewee, e.g. 'So, weather is not important in your holiday choice?'
Probe	Interviewer is asking for an explanation of the last statement, e.g. 'Why is the weather not important in your holiday choice?'
Back tracking	Interviewer returns to a point made earlier by interviewee to get further information/explanation e.g. 'You said before, you only took holidays in the UK up to age 23……'
New topic	Interviewer moves on to a new topic e.g. 'OK I think we have covered that topic. Where else did you take holidays in your 30s?'

■ Sample size

Unlike most quantitative research, it is not always necessary to decide in advance how many respondents to include when conducting qualitative research involving interviews. Instead, the approach is to use what is regarded as a 'sufficient number'. But how will you know when you have reached a sufficient number of respondents? Many texts (see e.g. Veal 2011, Long, 2007) suggest that this is when you have reached 'saturation point'. This is when you find that you are not been told anything that is new and different from earlier interviews (Glaser and Strauss, 1967; Long, 2007). However, in practice this point may be difficult to spot, and you may need to interview three or more 'extra' respondents before you realise that what you are being told is what you have been told before. The total size of your respondent group should also be linked to the aim of your research and the specific role or job of the respondents. For example, let us assume that your research aim is to investigate service quality and customer satisfaction in four-star hotel restaurants. You have decided to interview both waiters and guests and compare and contrast the results from each group. You could decide that you will do this in two hotel restaurants and interview five waiters in each and also five guests/restaurant customers in each. This will mean a total of twenty interviews – ten with waiting staff and ten with guests. Not only will you be able to compare the results from each group: waiters and guests, but also you have two locations to compare and contrast. This comparison work will become a key area of the analysis of your results.

You could interview just one member of staff and one guest in each restaurant. This would be only four interviews and of course would be far easier and less time consuming. Nevertheless, there are problems with just a small number of respondents. It is possible with such a small number that, for example, one of your waiters is just recovering from a bout of flu when you interview him and he does not respond as he would under normal circumstances, or possibly one of the guests was in an unusual hurry when dining and regarded the service as far too slow. 'Sufficient numbers' of respondents should offset these types of problem. Having set an ideal number of respondents, if you continue to be told new things then you can continue interviewing beyond your pre-set number, but you need to be aware of time restraints, not just for the actual interviewing, but of particular significance, the transcribing and later the analysis. Whatever the context, when interviewing, you will need to make a decision on when to stop and this may be when you are still gaining new data. However, you will need to be able to argue that the new material is relatively limited, and each successive interview largely provides responses that you have already been given in earlier interviews.

A summary of key factors in conducting a successful interview are indicated below, as well as what to avoid in an interview context.

How to conduct a successful interview

- Be polite
- Be empathetic
- Be confident
- Show interest in your interviewee's comments
- Encourage respondents
- Keep the interview flowing
- Keep the interview on track

- Watch the body language of your interviewee in relation to what they say
- Allow time for responses to be given, especially to complex questions
- Be flexible
- Expect the unexpected

What to avoid in an interview

- Do not be rude
- Do not be arrogant
- Do not treat it as if it was a job interview or test
- Do not cut respondents off sharply when they are speaking
- Do not judge or criticises their responses
- Do not be over-cautious
- Do not be over-confident
- Do not allow very long pauses in your discussion with the interviewee
- Do not stick rigidly to your interview schedule
- Do not allow lengthy discussion about topics which are not relevant to your research
- Do not look bored at your respondent's comment.

Case Study 5.2 is based on the work of a postgraduate student who used interviews as her main research technique. The case study refers to the first draft of interview questions. For the part of the research discussed below the student intended to interview senior managers of an airline, who were based in two different locations, and the focus of the research was on competition with other airlines.

The student prepared thirty-one questions for an interview intended to last approximately one hour. The initial questions were demographic (see below questions 1-5). Several of the questions were an attempt to ask respondents to conduct a SWOT analysis (Strengths, Weaknesses, Opportunities and Threats) of the airline and a selection of these questions is shown below, with my comments in italics provided as feedback to the student.

Case Study 5.2: Student's first draft of interview schedule with supervisor comments (comments in italics)

Interview Schedule

1 Name, Surname

Will you need both first and surname? Have you thought how you will present results and considered ethical issues, and the possibility of the use of pseudonyms?

2 Position

This is probably useful information, but have you considered precisely why in terms of your aim/objectives?

3 Nationality

This is probably important, but again have you considered why you need this?

4 How many years have you worked at X Airlines in Country A/Country B?

This appears to be a good relevant question

5 How many years have you worked in the aviation industry?

This appears to be a good relevant question

6 What do you consider are your key strategic resources? Fares, staff, service or what else?

An open-ended question would be better. If your respondent does not answer, then you could prompt them with the second part of the question

7 What is your current strategy? (probes: operations, marketing – brand)?

Although the probes appear to be relevant, will respondents understand what you mean by strategy here? Strategy in relation to what? A particular region or country, a particular theme/topic?

8 What is your competitive advantage (in addition to the positive attitude of your staff)?

Once again, an open-ended question would be better. You are providing respondents with one possible answer – which they may not have considered important anyway, or it may be, that what you have indicated about staff attitude is the response they would have provided!

9 What is the market share of the airline in Country X* and Country Y*

This is a factual question – will they know the answer? It could be better to ask: Is the market share adequate? And follow this up with: Could it be improved and how?

10 What proportion of your budget is spent on marketing? If you do not know the proportion, how much in money terms? Is it the biggest part of your budget?

This is a complex question- it appears to be two/three questions in one at the moment. Will respondents know how much money? Why do you need to know if it is the biggest part of the budget – is this linked to your aim and objectives?

11 Why did you decide to enter the market in country A*? What is more important, capacity or demand in the market?

This is at least two questions, so it is complex and ambiguous, and are the topics directly linked?

12 How do you segment your customer and do you use any criteria?

A good topic, but this is two questions in one and do you really want to find out what criteria they use, rather a Yes/No answer?

13 Why did you decide to use City Z* as a connecting hub? Who decided on route planning and flight frequency?

Once again (at least) two questions in one so it is too complex and possibly ambiguous.

14 How do local factors in Country B* affect flights? Is seasonality important?

Again two questions in one! I am not sure how the two topics are linked. However the seasonality factor could merit a question on its own – but you need to check your aims/objectives.

15 How do you regard economic conditions in Country A* in relation to the Asian market?

This question is not particularly clear. What does 'how do you regard economic conditions mean' and what is the link, you are concerned about, between the conditions in Country A and the Asian market?

- ***Some details have been changed, to preserve anonymity of the student.**

Focus groups

Focus groups are a fairly common technique for those conducting qualitative research in the social sciences (Krueger, 2004), although they do not appear to be used as much in the tourism and related areas as in other social science subjects. However, they have a number of advantages over other qualitative methods, as well as disadvantages, as discussed below.

Focus groups are similar to interviews but involve groups of respondents rather than just one. Instead of a one-to-one, face-to-face interview, the researcher acts as the facilitator of a number of respondents in a group context. The researcher may in fact ask precisely the same questions as she or he would in an interview situation. The main difference will be that there are, for example, ten respondents, sitting together in the group with the researcher, all potentially with comments in response to a question. The members of a focus group usually have something in common, although it is best if they have not met before, or they do not know each other well (Krueger, 2004). It is better that they have not met before because if they have, the behaviour of the group may not be 'normal'. In other words, some respondents may have previously developed antipathy to others in the group or conversely may want to agree with those they admire. However, what the members have in common may be involvement in a particular activity (e.g. attendees at a festival or taking part in a specific sport) or share certain demographic/spatial/temporal characteristics such as 'Afro-Caribbean women in the age range 20-25 living in a South London suburb in August 2022'. The members of a focus group will therefore be a sample of a specific group – in this case a purposive sample.

Ideally, these groups are between six and twelve respondents in size. The reason for this is that anything less than six tends to result in relatively limited amount of data being gathered (Krueger, 1994). Also, in excess of twelve participants, can lead to the groups being unwieldy with too many members trying to speak at the same time (Krueger, 1994). Over time this leads to only a few respondents contributing regularly during the discussion.

The context

Where and when the focus group takes place – the context – can be very important in terms of getting useful responses from respondents. Conducting a focus group in the changing room after a sports team have been heavily defeated is not likely to lead to positive comments about the team's chances of future success! However, asking a group of diners for their views on service or food quality, immediately after they have finished a meal at a highly regarded restaurant, could be a very good time and place to conduct focus group research. A key message is that if the focus group is related to an event, then the closest in time and space to the event that you conduct the group, the better the response to your questions should be.

What may put off some students from organising and conducting a focus group is that it requires certain skills to ensure that it goes well and produces useful data. A student may believe that they lack the skill or confidence to enable a discussion to take place, to ensure as many of those present are fully involved, does not become over-heated with a small number of dominating voices, possibly go off at a tangent, or probably worst of all, fail to get anyone in the focus group to say anything of importance.

The co-ordinator's role

The coordinator of a focus group has to be both a leader and a facilitator of discussion. Undertaking this role is arguably more difficult than conducting a face-to-face interview with just one respondent. In addition to conducting the focus group, the researcher needs to devise a way of recording what is said. It will be very difficult for the researcher to act as leader/facilitator and also manually record what is being said by respondents. Notes can be taken by the researcher, but this will be a difficult task if the researcher is also asking a number of questions and keeping the focus group discussion going. There are a number of possible solutions. If you are working alone, then an audio recording using a tape recorder, dictaphone or a mobile/cell phones (many have built in microphones) is probably the best approach. A video recording can be made, although if a real video camera is used this is more likely to interfere with participants than an audio recorder.

Recording the group

Whatever recording approach – audio or video – is used, you should be aware that some people 'perform' if they know that they are being recorded. This probably means that such people are not providing entirely accurate or truthful comments – they may say things just 'for effect' in order to provoke others. In a social setting, some people like to behave in such a way that they get others to support what they say and this can occur within a focus group. This may reduce the reliability of your results, as the group member who is seeking support for their views may be doing this not because they believe what they are saying, but merely to be socially accepted.

If possible, the preferred way to run a focus group is to have a colleague/friend attend the group and record manually on paper, the responses made. This person can also write down how comments/answers were made (e.g. 'strongly', 'defensively'), not just what was said. This person can also comment on body language and factors such as ongoing involvement of certain individuals, to indicate if they are dominant voices, or only occasional responders. At the same time as your colleague is manually recording aspects of the focus group, you can also be using an electronic audio or video device. After the focus group you will then be in a position to listen to, or watch your electronic recording device, and compare this with the hand-written comments. This should enable a greater possibility of better analysis of your results. Case Study 5.3 presents a summary of two focus groups I conducted in connection with a music festival in Devon, UK. One focus group was held during the Festival, the other a few weeks after.

Case Study 5.3: Two focus groups used in connection with a music festival

Background

This focus group research was part of wider research into attitudes to the Sidmouth International Folk Festival and involved both visitors and locals. There were two focus groups. The first was intended to involve local residents only and took place during the actual week in August that the Festival occurred. The second group involving 'retailers/traders' was scheduled for several weeks after the Festival itself, in October, as the 'retailers/traders' were working at the Festival. The research was financially supported by the local council. As a result, the council also organised the focus groups, including recruiting the member of both groups, and twelve participants were invited for each group.

Group 1

This was held on the Wednesday morning of the Festival week – the Festival runs from Friday evening to the Saturday evening eight days later.

- Despite inviting twelve participants, only six came to this group. They could be best described as 'ordinary locals', although they were in the age range forty to seventy, there were four men and two women.
- Despite the small number of participants, there was generally a good dynamic in the group.
- There were two dominant voices, both male.
- In addition, there was one 'whinger' who spent most of the approximately one hour complaining about almost everything and stayed behind on his own to have an extra half hour to continue moaning!

Group 2

This was held on a Thursday morning in October of the same year following the Festival in August.

- The intention was that the group should be made up of twelve 'local traders' who had been invited to attend by the local council. Eleven came to the focus group. Seven of those invited who came were appropriate to this group. There were three hoteliers, two shopkeepers and two sport club members (the sports club facilities had been used during the Festival for events and parking).
- There were four members of the group who had been invited by the local council to attend Group 1 – a Catholic priest and three local Morris dancers!

- As a result of the mixture of participants, there was a strange dynamic. The non-traders clearly felt that they did not have much to contribute, but the priest made many, largely useful, comments. However, two of the hoteliers (who, it became clear at the end of the focus group meeting, had recently divorced each other), verbally attacked each other, but also hid behind many of the questions as they involved 'commercially sensitive information' they claimed, so would not provide detailed responses.
- Despite the above, significant new information and insights were gained.

I have also used focus groups as a part of research which involved a number of different techniques and a range of respondents. In one example of this, a colleague and I were asked to provide information to help a local council in New Zealand plan for new and emerging sporting activities in the region, one of which was mountain biking. Initially we conducted a questionnaire survey of members of the local mountain biking club. This questionnaire asked respondents to indicate if they were willing to take part in a follow-up focus group. Twenty-three respondents to the questionnaire indicated that they were willing to do this, and they were subsequently invited to attend one of two focus groups held at the local council offices. Case Study 5.4 discusses the conduct and results of the two focus groups that were held on the topic of mountain biking.

Case Study 5.4: Mountain biking focus groups, New Zealand

The setting was a medium size New Zealand town in the late 1990s. The traditional sports that New Zealand was known for were rugby, cricket and netball. Mountain biking had been growing for a decade or so, but was still small scale. The researchers were assisting the local council to prepare a leisure/recreation plan.

The focus groups followed on from a questionnaire survey of the members of the local town mountain biking club in which there were eighty respondents.

Twenty three out of the eighty had indicated in the questionnaire responses that they were willing to attend a focus group.

Two consecutive focus groups were organised with the help of the local council. Twelve mountain bike club members were invited to one group and eleven to the other. However only sixteen turned up – one group had twelve participants, the other only four.

There was a gender imbalance in both groups – only one person out of four in the smaller group and four out of twelve in the larger group were women. However, this closely mirrored the nature of mountain biking participants at the time.

There were three dominant voices in the larger group and only half made significant contributions. In the smaller group, the only female in the group was very much the dominant voice.

Despite the issues above, respondents were able to provide explanation to some of the earlier questionnaire responses. They also gave new information, leading to a better understanding of who mountain bikers actually were and their needs in relation to other user groups.

One other interesting result of these focus groups was that this form of direct contact with the mountain bikers meant that respondents felt that the focus group organisers/researchers could help solve their problems!

The case studies of the use of focus groups provide an indication of lessons that can be learned. A key factor is that, if possible, do not let others organise your focus groups for you, or you may find yourself in the position I did in relation to my Group 2 for the Sidmouth International Festival. Also, try to ensure you have sufficient respondents in each group – unlike one of my groups of mountain bikers with only four attendees. In addition, try to ensure that your group members have something in common in terms of aspects that you wish to research, but do not know each other before they meet in the group. My case studies provide some evidence for there being a number of advantages and disadvantages of focus groups, particularly in relation to other qualitative research techniques such as interviews. These advantages and disadvantages are summarised below.

Advantages of focus groups

- They involve more respondents than a single interview, so can save much time
- Several consecutive groups can be held on the same topic leading to a relatively large 'sample'
- Respondents can 'spark each other off' as one respondent provokes a response from another, which may be similar or different but would not have occurred in a one-to-one interview
- They can be used as a 'stand-alone' technique, or prior to other research (such as a questionnaire) or following on from other research

Disadvantages of focus groups

- There are likely to be dominant voices and quiet voices, but the dominant voices do not always provide useful data
- There will usually be 'unheard' views
- Groups can be too large so that some members say little or nothing
- Groups can be too small, so a range of views is not provided
- The facilitator needs skill to make sure discussion flows and produces useful data. If not it could descend into...'what did you watch on television last night?'

Writing this section in your study

The discussion of your overall research methodology and specific techniques would be best located in your study, in what is likely to be a chapter titled 'Methodology'. This could follow on directly from a chapter reviewing the literature – this is the convention in most dissertations or theses, particularly those using a quantitative research approach. However, in a qualitative study you may place the Methodology chapter before the literature review. Whatever is the case, the techniques presented in this chapter, would be best discussed in the 'Methodology' chapter, but following on from a discussion of research philosophy.

The discussion will be based on the specific technique, or techniques that you have used. You will need to provide a rationale for the technique(s) used. This will require reference to literature. This discussion will need to show precisely what you intended your research findings would be when using the chosen echnique(s). It should also consider other possible ways that you could have gathered data and why you did not use these but selected the particular technique(s) you did use – this is part of the rationale for the choice of the technique(s).

The creation of the specific questions you have used will need to be explained, particularly in terms of the sources, be it academic or other sources. You will need to provide a rationale for the particular questions you have used. You will also need to indicate how you put the questions into the order presented to respondents and the rationale for this.

You will almost certainly have carried out some form of pre-testing of your technique or perhaps a pilot study. You will need to explain the precise intentions of the use of the pre-test or pilot, and what, if any, implications of the outcomes of this there were for your 'main research'. This will not require a lengthy discussion of the results of the pre-test or pilot, but, if you made some changes, you can write something to this effect: 'On the basis of the results of the pilot study, one demographic question and two of the questions on the main research topic were modified.'

How you used the specific technique(s) in the 'main research' will be an important part of this discussion. If you had particular problems, these should be referred to as well as the way in you dealt with these issues. If you used, for example, interviews or focus groups you will need to indicate and discuss how your group of respondents was selected. If we assume that you used a purposive sample or a convenience sample, you will need to discuss the nature of this form of sampling and its advantages and disadvantages, and this will require reference to literature.

If you used observation, you will need to indicate whether this was participant observation or non-participant observation. Also, you will need to inform the reader whether the research was overt or covert. The rationale for the choice of 'overt' or 'covert' will need to be made, particularly if covert observation was used. How you actually conducted your observation will need to be discussed as well. In terms of design and conduct of the observation, you will need to refer to literature on these areas to describe and explain what you did.

Ethical issues will almost certainly have to be discussed if you used covert observation and are very likely to be a feature of your discussion if you used overt observation. For whatever type of qualitative technique(s) you used, explaining how you obtained 'informed consent' from participants will be important.

Student activities

1 What are the main differences between participant and non-participant observation?

2 Under what circumstances would you use *overt* observational research? Under what circumstances would you use *covert* observational research?

3 What are the main similarities and differences between interviews and focus groups?

4 Under what circumstances would you use focus groups rather than interviews?

5 A student I supervised evaluated the use of the following qualitative research techniques: open-ended questionnaires, focus groups and semi-structured interviews. He used the following criteria in this evaluation:

- *effectiveness* (the ability of the technique to produce useful responses),
- *efficiency* (the amount of data generated),
- *depth and detail* (the quality of the data produced)
- *uniqueness (t*he ability to produce data that no other method produced).

His evaluation indicated that using semi-structured interviews was the best research technique of the three. Why do you think this was the case?

6 Mixed Methods Research

Introduction

Although it has been conventional in the past to conduct research in just one paradigm, increasingly researchers have used a combination of both research paradigms in what is known as mixed methods research. This means that research could initially be qualitative and then in a follow-up stage be quantitative, or it could be initially quantitative and then in the next sequence be qualitative. In some large-scale research, it has also been the case that a qualitative approach is being used consecutively with a quantitative phase of research, by the same research project team.

There are a number of issues with the use of mixed methods. Probably the most important is that for many researchers using both paradigms together, is viewed as not philosophically appropriate. Such researchers argue that if the world is viewed as being one where objective truths can be found, how is this compatible with the world view that there is no such thing as objective truths and the world should be considered in relation to individual interpretations and is therefore subjective? Although this view has held sway for many years and still has many supporters, there are researchers across many of the fields of social science, including tourism and related subjects, that are willing to use, and in fact advocate, a combination of methods. Such researchers may, for example, use interviews initially and follow this up with a questionnaire survey. Despite these changes and the espousal of mixed methods by a number of researchers, you should be aware that many researchers, and this may include your supervisor and examiner, may not support this use of the combined approach. If you are thinking of using a mixed methods approach, it is probably a good idea to consult your supervisor.

Mixed methods research

The use of mixed methods, although not new, is a fairly recent phenomenon in tourism research (Khoo-Lattimore et al., 2019; Mason et al., 2010; Seakhoa-King et al., 2021) and there has been relatively little research employing mixed methods compared with the use of just one research philosophy, be it either quantitative or qualitative (Khoo-Latimore et al., 2019).

Early research in tourism, going back to at least the 1960s, was largely quantitative (Mason, 2017). The starting point for much early quantitative research was government or company documents, and this approach was frequently supported by the primary research technique of the questionnaire survey (discussed in Chapter 4), which usually sought generalisable results from a sample of respondents who provided mainly short answers, via closed-ended questions, that could be statistically analysed. Reaction against this quantitative approach in tourism research was appearing during the 1970s in the work of sociologists and anthropologists such as Cohen (1972) and Dann (1977). For such researchers, the use of interviews and focus groups was relatively common (Lincoln and Guba, 1985), as they were interested in views and opinions and, as we have noted in Chapter 5, these can best be expressed in depth and detail using the words of respondents themselves, rather than answers generated via the use of author-designed, closed-ended questions, such as those on a questionnaire, it was argued (Creswell, 2018; Mason, 2014).

The discussion above outlines the nature of the two predominant research traditions used in tourism research over the past fifty years, but a third approach has been applied during the last thirty years or so. This, mixed methods research, is less well known than either of the two other approaches, partly because it is relatively new (Teddlie and Tashakkori, 2009). As Teddlie and Tashakkori, have indicated it presents an alternative to the quantitative and qualitative traditions by:

> 'advocating the use of whatever methodological tools are required to answer the research questions under study' (2009: 6).

In terms of research design, Tashakkori and Teddlie (2003) defined mixed methods research as where both quantitative and qualitative approaches are used in types of questions, research methods, data collection and analysis, while Tashakkori and Creswell, (2007) also stressed the links between the two main research traditions, when defining mixed methods as research where the investigator collects and analyses, and integrates findings, using both qualitative and quantitative approaches.

However, it should not be forgotten that the use of mixed methods is frequently viewed as controversial (Teddlie and Tashakkori, 2009; Timans, Wouters and Heilbron, 2019) and for some researchers combining quantitative and qualitative design, data collection and analytical processes is regarded as inappropriate, as the underlying philosophies are incompatible (Guba 1987; Smith and Heshusius, 1987).

Research philosophies

The research philosophy of *positivism*, as indicated in Chapter 3, and until fairly recently regarded as the dominant approach in tourism research (Khoo-Lattimore et al., 2019; Riley and Love, 2000; Walle, 1997), views reality as singular and independent of the researcher (Creswell 2018; Decrop 1999; Lincoln and Guba, 1985). It also regards reality as something which can be measured objectively, for example, by using a quantitative research instrument (Teddlie and Tashakkori, 2009; Davies, 2003). However, the research philosophy of *interpretivism* rejects the idea of a single objectively measurable reality, positing instead the existence of subjective, multiple constructed realities (Creswell 2018; Davies 2003; Lincoln and Guba 1985). The individuals involved in the research situation here, including the researcher, those being investigated, and the reader or audience interpreting the research, are regarded as the creators of these realities (Creswell, 2018). Interpretivists contend that multiple realities can only be captured through the employment of less rigid data collection techniques than generally used in quantitative research (Creswell and Plano Clark, 2007; Bernard, 2000).

The research philosophies of positivism and interpretivism also differ in their assumptions about the relationship between reality and the researcher (Creswell, 2018). Interpretivism has its roots in 'Verstehen' or the empathetic understanding tradition of the sociologist Max Weber, where it is argued that to understand peoples' behaviour, researchers need to put themselves in the place of the enquiry subjects (Jennings, 2011). Therefore, interpretivist researchers try to minimise the distance between themselves and respondents (Creswell, 2018). So, the interpretivist tries to get inside the respondents' minds and see the world from their view (Jennings, 2011). Interpretivists have been described as observers 'from the inside', the term used is '*emic*' (Teddlie and Tashakkori, 2009; Phillimore and Goodson, 2004; Walle, 1997). However, positivist researchers try to provide an independent, outsider's account of the research process, known as an '*etic*' approach (Teddlie and Tashakkori, 2009; Phillimore and Goodson, 2004; Walle, 1997) to ensure that the researcher's biases are excluded.

Although, the two research philosophies may seem incompatible, as stated above they are being increasingly combined in the social sciences, including in tourism (Timans et al., 2019; Mason et al., 2010; Seakhoa-King et al., 2021) and what makes it possible to use them together in one study, is the underlying philosophy of the majority of mixed methods research, that of '*pragmatism*' (Teddlie and Tashakkori, 2009). Pragmatism has been defined as a research approach that debunks and demystifies concepts such as 'truth' and 'reality' and focuses instead on 'what works' as the 'truth' regarding the concepts under investigation. Pragmatism rejects the either/ or choices of the battles between paradigms and advocates the use of mixed methods in research (Tashakkori and Teddlie 2003: 713).

Despite mixed methods being a fairly new research approach, a number of different types can be identified. Creswell and Plano Clark (2007) identified twelve different classification systems in mixed methods, drawn from several fields in the social sciences, including health, education and behavioural studies. Creswell (2009) also suggested there are six types of mixed methods research. There are two major groupings in this sixfold typology, based on whether different research approaches are used sequentially or concurrently (Creswell 2009). In relation to sequential research, one approach begins with qualitative research and is then followed by a quantitative phase, and this is termed *sequential exploratory research* (Creswell, 2009). Research starting with a quantitative approach, followed by a qualitative phase is termed *sequential explanatory research* (Creswell 2009). The third type of sequential research uses a specific theoretical perspective from the beginning, which shapes the direction of the research, and the sequence of the two-phase research can begin with either a quantitative or a qualitative approach and Creswell terms this *sequential transformative research.*

In terms of the second major grouping of mixed methods research, Creswell indicates there is *concurrent triangulation*, where both qualitative and quantitative data are collected concurrently. In this approach, the intention is to give equal weighting to each of the qualitative and quantitative approaches. However, in contrast, in the *concurrent embedded approach*, there is one predominant approach, and the secondary supporting approach (quantitative or qualitative) is embedded within the main approach. Finally, in this six-fold typology, there is the *transformative concurrent approach* where the research is guided by the use of a specific theory in the collection of both quantitative and qualitative data, concurrently (Creswell 2009).

Doing mixed methods research

In addition to the philosophical issues indicated above, for some researchers, it is the practicality of doing mixed methods research that is a regarded as problematic. It is perceived that it is likely to take much longer to conduct mixed methods research than using just one approach. This is frequently the view of students and often supervisors, and clearly if you are using, for example, two different techniques you have to design each instrument, administer it and then analyse the results. Under these circumstances, you will probably believe that it could take up to twice as long to do this as using just one approach. However, this is not necessarily the case. If you use two techniques or approaches, then it may be possible to reduce the perceived time required and also make resource savings. For example, if conducting a questionnaire survey, you could also use a number of interviews, and these could be with some of those from your questionnaire sample – so you are not having to find new respondents. The purpose of the interviews would be to gain (further) explanation on the questionnaire responses and also additional comments relating to the research topic. So, you would hope to gain extra data compared with just using your questionnaire. Also in this example, the actual number of the interviews would not necessarily have to be as many as the total required if the interviews were your only research technique. Hence, there may not necessarily be a great need for extra time to conduct this type of mixed methods research and you may gain new insights into your research topic.

The reference to gaining new insights, in the previous paragraph, indicates a major advantage of using a mixed methods approach – you are able to compare the results from your different approaches. You will see later in the book, in chapters concerned with analysis, that being able to compare – to look for differences and similarities – is at the heart of good analysis and this process should be at the core of your research prject. If you were using, for example, just a questionnaire, you would be able to compare answers to individual questions by your whole sample of respondents and also compare these results with literature-based research projects on your topic. However, if, for example, you use interviews in addition to a survey, you could compare the results of the interviews with all the other sources of data that you have. The major reason for using interviews, in addition to your other approaches, would be to make your results more reliable. The process of using different techniques as discussed above is an example of *triangulation*.

Triangulation

This is the use of a variety of techniques (and can include methodologies) with the intention that your results are more reliable. The rationale, therefore, is to try to make your overall study more reliable, more credible and of better quality in an attempt to gain a higher mark.

Strictly speaking triangulation is a process that involves more than one technique or method – it does not have to involve three, as its name seems to imply, just more than one. You can have different techniques within one research paradigm. So, you could use, for example, observation and focus groups within qualitative research for your study. Alternately, you could use document analysis and a questionnaire survey within quantitative research. Whatever your particular single paradigm-based approach, the triangulation will enable you to compare the results from the different techniques used. Indeed, the rationale for the use of more than one technique is to be able to compare your results from the different techniques and this should make your results more reliable.

However, in relation to the mixed methods approach and triangulation, there is another important advantage. If for example you have used a questionnaire survey with closed-ended questions, then you could follow this up with the use of interviews. The interviews, which are within the context of the use of a qualitative paradigm, could be with a number of the respondents who have completed the questionnaire. The rationale for the use of interviews with this group of respondents would be to provide them with an opportunity to explain their closed-ended questionnaire responses and possibly add more comments on these. In this situation the respondents would be answering in their own words. It would then be possible to compare the questionnaire responses with the interview responses. Similarities and differences could be noted and would provide the basis for what would very probably be more detailed analysis than if just using one technique.

The mixed methods approach would also be feasible and probably useful in the context discussed next. In exploring a new topic, you may have decided to use initially a focus group, as you want to hear respondents' comments in their own words and another part of the rationale for the use of the focus group is that you are not certain what the key issues in relation to your topic will be. When you conduct an initial analysis of your focus group responses, this may produce a number of topics which you can then use to design and produce a questionnaire. The questionnaire can be used with a sample of respondents which will almost certainly be a larger group than the total number of your focus group respondents. Having conducted the questionnaire research, you will be able to compare the focus group and questionnaire results. You will also be in a strong position, when writing up your research methodology, as you will be able to make the important point to defend your research design that your questionnaire questions came directly from your initial focus group question responses.

Case Study 6.1 gives an example of the type of mixed methods research discussed in the paragraphs immediately above. It provides an example of the use of a quantitative approach – the use of a questionnaire survey – which preceded a qualitative approach, that of the use of interviews, and those involved in the interviews were some of those who had responded to the questionnaire survey.

Case Study 6.1: An example of mixed methods research in a Masters' dissertation.

My postgraduate degree was a purely research Masters, as I had originally begun a Master of Education (M.Ed.) course, which involved a number of taught modules and additionally a dissertation, which is similar to the model in which most Masters' students will be involved. However, I transferred from the M.Ed. to a Master of Philosophy (MPhil.). For this MPhil., I was only required to do a research dissertation rather than some taught courses, plus a dissertation. However as an M.Phil, the dissertation was required to be longer, at 50,000 words, than most Masters' courses, but not as long as a PhD thesis.

When it was originally planned (as part of my M.Ed. course), I had intended to conduct solely a questionnaire survey. My respondents were to be secondary school teachers and I was investigating the relationship between global development issues and tourism. However, once I had transferred to the MPhil, I realized I would need more data. My supervisor suggested that I should try to gather this using a different technique from the survey. So, my approach was to use interviews. However, I linked both techniques. I conducted the survey first and asked respondents to indicate if they would be willing to take part in a follow-up interview. The aim of this interview was partly to get further comment on, and explanation of, some of the survey responses. I also wanted to ask a number of additional questions. Nearly all of these were open-ended questions, to be answered in the respondents' own words. These were seeking their views on a number of topics related to my main theme of tourism and development. I recorded the interviews on tape and transcribed them before analysing, using both the question topics and the data themselves to create analytical categories.

Although I had used a mixed methods approach, I was not fully aware of this until my viva. As the only piece of work that I had conducted for my Masters' degree, it was a requirement at my university that I had a viva. This is an oral examination concerned solely with the dissertation/thesis. One of the examiners asked me a question about mixing paradigms and possible problems this could cause. I seem to remember that I answered a different question which was to do with 'mixing techniques' (interviews and questionnaires). The examiner was not impressed and followed up by asking why I had used two different techniques. Fortunately, I was able to provide, at least a partly convincing response, about the advantages and disadvantages of each technique and how the disadvantages of questionnaires could be offset by the advantage of interviews!

Case Study 6.2 gives another example of the approach also discussed in the paragraphs prior to Case Study 6.1, but here, a qualitative phase preceded a quantitative stage. In this case, which involved exploratory research, a variety of qualitative techniques were tested prior to the use of a questionnaire survey. The research involved asking tourists to indicate what they believe a quality tourism destination is. The Tourism Destination Quality (TDQ) study used both qualitative and quantitative techniques and involved a sequential exploratory mixed methods approach (Creswell, 2009)

Case Study 6.2: Sequential mixed methods research

This case study indicates the nature of the phases of research in the TDQ study, beginning with the exploratory research involved. It discusses how questions were designed, and the pre-testing and piloting of questions. It provides details on the main qualitative and subsequent quantitative phases of the research.

At the start of the research process for the TDQ study, two significant questions required answers. These were:

a) What type of research problem/issue was being investigated?
b) How should research be conducted once the nature of the research problem/issue had been clearly identified?

The response to Question (a) depended on whether this was a problem/issue that had been previously well-researched or a new/relatively un/under researched one. Once Question (a) had been answered, it would be possible to answer Question (b).

In relation to Question (a) a literature search indicated little was known about the issue of destination quality from the tourists' perspective; this was a new topic and hence the research was regarded as exploratory. As there was no known pre-existing literature on the issue, an attempt to gain insights into and a clearer understanding of it, would require results to be in-depth and in detail, it was decided. However, the intention of the research was not to just gain depth and detail, which would almost certainly mean obtaining views from a relatively small number of tourists, but also to gain a broader understanding from a larger population of tourists. Therefore, in relation to Question (a) above, attempting to gain depth and detail concerning the issue, but also a wider understanding of it, had implications for the conduct of the research, specifically in terms of the nature of the underlying philosophy and the data collection techniques to be used.

It was clear from what was required, that two philosophical and practical approaches to conducting the research would be required, but also that they would need to be arranged in sequence. Therefore, of particular importance, in terms of the practicalities of actually doing the research, a decision on how the two approaches would be sequenced was required. Below are the questions in chronological order that were created once the two initial questions, (a) and (b) above, were answered.

1. What research philosophy should be used in the first phase of research?
2. Which research techniques should be used in the first phase of research?
3. What specific questions should be asked in the first phase of research?
4. How should pre-testing take place in the first phase of research?
5. How should a pilot study be created and used in the first phase of research?
6. How should the results of the pilot study be used to help design the main research instrument in the first phase of research?
7. How should the research in the first phase be conducted?
8. How should results of the first phase of research be used in the second stage?
9. Which research philosophy should be used in the second phase?
10. Which research techniques should be used in the second phase?
11. What specific questions should be asked in the second phase?
12. How should a pilot study be created and used in the second phase?
13. How should the second stage of research be conducted?
14. How should the results from the second stage be analysed?
15. What were the overall findings of Phase 1 and Phase 2?
16. What conclusions could be drawn from the findings?
17. What recommendations (if any) could be made, based on the findings?

As is indicated in Case Study 6.2 you will need to analyse your results. The way in which you conduct analysis of mixed methods is to follow the approach of the analysis of each of the quantitative and qualitative techniques you have used. You may want to write up the analysis of each technique separately but will almost certainly want to have a section of your study which compares the results of each of your techniques/approaches. Chapters 8 and 9 present and discuss approaches to analysing quantitative and qualitative results, respectively. However Chapter 10 discusses the analysis processes specifically when using mixed methods.

Writing this chapter in your study

If you used a mixed methods approach, you will need to explain precisely why you did this. This should involve discussion of key terms including ontology and epistemology. Although you may not want to go into detail on the so-called 'paradigm wars' (Teddlie and Tashakkori, 2009) it would be sensible to indicate that you aware of these controversies. This could help with the rationale for the use of mixed methods that you will also need to provide. This rationale, which should discuss advantages and disadvantages of mixed methods, will need to be supported by literature.

As the use of mixed methods is still relatively new and not as typical as the use of just one research paradigm, and this is also the case for dissertations and theses, the rationale you provide will need to be particularly robust to convince your supervisor and examiner that you knew what you were doing, as well as why and how.

The conduct of the research will need to be described and particularly the actual sequencing and why you conducted the research in the order used. Any issues, including ethical issues, you had while conducting the research will need to be presented, as well as how you overcame any problems.

You will also need to provide details on the specific techniques you used and the rationale for whichever ones you employed. It is likely, although not necessarily the case, that you have used a questionnaire as part of the mixed methods research. If you have, then you should make use of the points indicated in Figure 4.9 which discusses how you should prepare and conduct your questionnaire. Other points in the section 'Writing this section in your study' at the end of Chapter 4 may be useful as well.

There are several possible qualitative techniques that you may have used. You should read the section at the end of Chapter 5 'Writing this section in your study' to assist, whatever technique(s) you actually used.

Student activities

1 What do you understand by the term 'paradigm wars'?

2 Create a two column table and head the left hand column 'Advantages of mixed methods research' and head the right column 'Disadvantages of mixed methods research?' When completed, indicate which of the two columns has the more important factors.

3 What do you understand by the concept of triangulation? Under what circumstances would you make use of triangulation? What are the advantages of triangulation?

4 With reference to Case Study 6.1, what are advantages and disadvantages of following up a questionnaire survey with interviews that involve some of the participants that responded to the questionnaire?

5 With reference to Case Study 6.2, what benefits are there to following up interviews with a questionnaire survey?

7 Presentation of Results

Introduction

There are many different ways of presenting your results. If your study is of a quantitative nature, it is conventional to use actual totals of responses, but more commonly data are presented in various statistical ways as percentages, figures, tables, graphs, often using computer software packages.

If you have used a qualitative approach, it is less likely that you will use graphs and tables, but this is still possible. However, it is more likely that you will use actual direct quotations from your respondents as well as summaries of responses. Direct quotations are often referred to as 'verbatims'. A verbatim statement (a 'verbatim' for short, although grammatically incorrect!) contains the precise words of one of your respondents. What distinguishes the qualitative presentation of results from the quantitative is that it is largely in the form of words. However, some summary tables and figures may be used. There are also computer tools that have been specifically designed to analyse responses which are in the form of words.

Whether your results are of the quantitative or qualitative type, there are no set rules on how you present your results. However, what should guide you are as follows:

- Trying to be as clear as possible
- Making sure that your results can be understood
- Ensuring that the table/graph/figure you use to show your result is at least as good as if not better than using words. By the way, there is no point in creating a table or graph to show the results and then writing in words precisely what the graph/table shows. It is better to use one or the other. If you have used a diagram, you can then write for example, 'As Figure 1 shows…'. Your commentary on Figure 1 would then be a brief summary indicating the more/less significant aspects.
- Try to ensure your presentation is attractive. This is the least important of these points, as having attractively presented results that could

have been expressed in words, or in other ways than you have used, will not gain you any extra marks. However, poorly presented results will certainly not gain you marks either and will give the impression to your examiner that you care little about your work.

Presenting quantitative results

Assuming you have used a questionnaire, before conducting the analysis of this, you will need to prepare and present your results. Although you could include all your completed questionnaires and ask readers to look at them, this is not an appropriate approach. The main reason being that you will not have done anything to help the reader understand the results if they are presented in this way. You have merely compiled them. The results need to be prepared before being presented. The most likely way that you will do this is to summarise them. You can do this by using simple statistical techniques, but also by summarising comments respondents have made to any open-ended questions, or if they have been asked to explain answers in their own word.

There are a number of relatively simple and straightforward ways to present results which are in numeric form. It is often worthwhile presenting the actual figures, as in, for example, the total number of respondents in your sample, or the number of respondents in a particular category. However, it is frequently preferable and desirable to convert actual totals to percentages. The major advantage of converting actual figures to percentages, is that it is possible to compare results from different size response groups or samples. For example, you may have conducted a survey, and obtained a usable sample, of 45 respondents at one location and 57 at another location. If you had 30 respondents giving responses in a specific category at the first location and 38 in this same category in the second location, how can you directly compare the results? The numbers giving the same response in different locations appear different, until they are converted to percentages. Hence 30/45 in your first sample is 66% and 38/57 in your second sample is also 66%. So, despite different sample sizes, the percentage in each sample is the same and this may be a particularly important result for you to comment on and discuss in your analysis.

Converting responses to percentages is a relatively straightforward task and can be done manually, particularly if your sample size is relatively small. It is useful with closed-ended questions when there are, for example, only two response categories, such as 'Yes' and 'No'. However, it is also possible to use with several closed-ended response categories. For example, you may have a number of age categories:

21-30
31-40
41-50
51-60
61+

Percentages can be provided for the total in each of these age categories. Other demographic factors such as ethnicity, education level and income level are commonly shown in surveys as percentages. Converting your responses of this type to percentages also gives you information on whether your sample is representative, reliable and valid, particularly if you are able to compare the sample you have obtained with any published work you have used in the preparation of your survey.

Percentages can also be used in relation to questions about the substantive topic you are researching. If you have made use of published work in preparing your survey, it may be possible to compare your results directly with the published work. Figure 7.1 below shows a question I used in relation to research at Wellington Zoo, New Zealand. This question has been presented earlier in the book, but here the percentage responses to the question have been added in.

Figure 7.1: An example of a checklist question: Motivations for visiting Wellington Zoo, results with percentages shown

Family outing	44%
To see the animals	28%
A day out	10%
School/other group	16%
Visit with a friend	1%
Other	1%

Note that the total (adding together all the percentages in the right-hand column) is 100%. Also note that the 'Other' category has only 1% in it, strongly suggesting that the categories I have chosen are valid. It would a relatively easy task to convert this table of percentages to a pie chart or a bar chart (see below for a discussion of creating pie charts).

Figure 7.2 shows another question I used in relation to research at Wellington Zoo, New Zealand. This question has also been presented earlier in the book, but the percentage responses to the question have been added in here. Note that, in this case, each row has percentage totals adding up to 100%. These results clearly show important differences, as indicated by respondents to the role of the zoo, with a very large proportion of those responding indicating that they agree with the importance of the education role of the zoo, and a small majority disagreeing with the statement relating to the tourism role of the zoo.

Figure 7.2: An example of Likert scale statements used for research at Wellington Zoo, New Zealand

'It is an organisation ...	Agree	Disagree	'Neither Agree or Disagree'
that mainly breeds rare animals	63	15	22
set up mainly to educate people	94	3	3
set up mainly as a tourist attraction	32	53	15
set up mainly to conserve animals	54	24	22

The question in Figure 7.2 uses statements and a Likert scale and indicates that it is also possible to give percentage responses in each category which in this case are 'Agree', Disagree' and 'Neither Agree, nor Disagree'. This question actually shows 'aggregated responses' for the 'Agree' and 'Disagree' categories. They have been achieved by combining 'Strongly Agree' and 'Agree' to create a category '*Agree*' and combining 'Strongly Disagree' and 'Disagree' into '*Disagree*'.

To illustrate how we create percentages, if we take an example with 200 completed responses using a Likert scale with five categories ranging from 'Strongly Disagree' to 'Strongly Agree' and the total number in each is as follows:

Strongly Disagree=16, Disagree=24, Neither=10, Agree=128, Strongly Agree= 22

then this can be converted to percentages in each category. The results of this conversion to percentages is as follows:

Strongly Disagree=8%, Disagree=12%, Neither=5%, Agree=64%, Strongly Agree=11%.

(Note that the percentages add up to 100).

It is conventional to convert the results in a Likert scale to obtain a mean score. The main reason for this is that the mean scores from several Likert scale statement questions can easily be compared. The conventional way to create mean score for a Likert scale is shown in Figure 7.3 below.

Figure 7.3: Examples of Likert Scale mean values

Example 1 Using the five-fold classification and the example above, the mean is calculated in the following way:

Strongly Disagree (1)	Disagree (2)	'Neither' (3)	Agree (4)	Strongly Agree (5)
16	24	10	128	22

Multiply the numeric value of each category by the number in each category:

Strongly Disagree (1)	Disagree (2)	'Neither' (3)	Agree (4)	Strongly Agree (5)
16	48	30	512	110

Total of all (add up each score) 16+48+30+512+110 = 716

Divide the total (716) by the number of respondents (200) = 3.58. The mean =3.58.

Example 2

Strongly Disagree (1)	Disagree (2)	'Neither' (3)	Agree(4)	Strongly Agree (5)
8	16	10	94	72

Multiply the numeric value of each category by the number in each category:

Strongly Disagree (1)	Disagree (2)	'Neither' (3)	Agree (4)	Strongly Agree (5)
8	32	30	376	360

Total of all 8+32+30+376+360 = 806

Divide the total (806) by the number of respondents (200) = 4.03. The mean = 4.03.

Example 3

Strongly Disagree (1)	Disagree (2)	'Neither' (3)	Agree (4)	Strongly Agree (5)
144	32	8	10	6

Multiply the numeric value of each category by the number in each category:

Strongly Disagree (1)	Disagree (2)	'Neither' (3)	Agree(4)	Strongly Agree (5)
144	64	24	40	30

Total of all 144+64+24+40 +30 = 302

Divide the total (302) by the number of respondents (200) = 1.51. The mean =1.51.

But what do these mean values actually indicate? A mean score of 3.58 (as in Example 1) indicates that respondents are to some extent in agreement with the statement. The mean, however, is fairly close to the mid-point value (3). However, relatively few respondents indicated 'strongly agree' in relation to the statement. Also, there is a small, but important minority (20%, 40/200) who do not agree with the statement (this has been achieved by combining or aggregating the 'Strongly Disagree' and 'Disagree' categories). So, taking together the 20% who 'do not agree' with the 11% (22/200) who 'strongly agree', reveals that, despite 64% (128) in the 'agree' category, the mean value is only 3.58.

In Example 2, a large number of respondents have answered in both categories 'Agree' and 'Strongly Agree'. Very few respondents have answered in either of the 'Disagree' or 'Strongly Disagree' categories. Additionally, there are few respondents in the 'Neither Agree, nor Disagree' category. Consequently, the mean is over 4 (remember this out of a possible maximum mean of 5).

In Example 3, a large number of respondents have answered in both categories 'Strongly Disagree' and 'Disagree'. Very few respondents have answered in either of the 'Strongly Agree' or 'Agree' categories. Additionally, there are few respondents in the 'Neither Agree nor Disagree' category. Consequently, the mean is low, being almost half-way between 1 and 2 -remember that the minimum mean is 1.

In addition to converting actual results to percentages, there are a number of different charts, tables, graphs and figures that can be used to present your results. A bar chart is shown below (see Figure 7. 4).

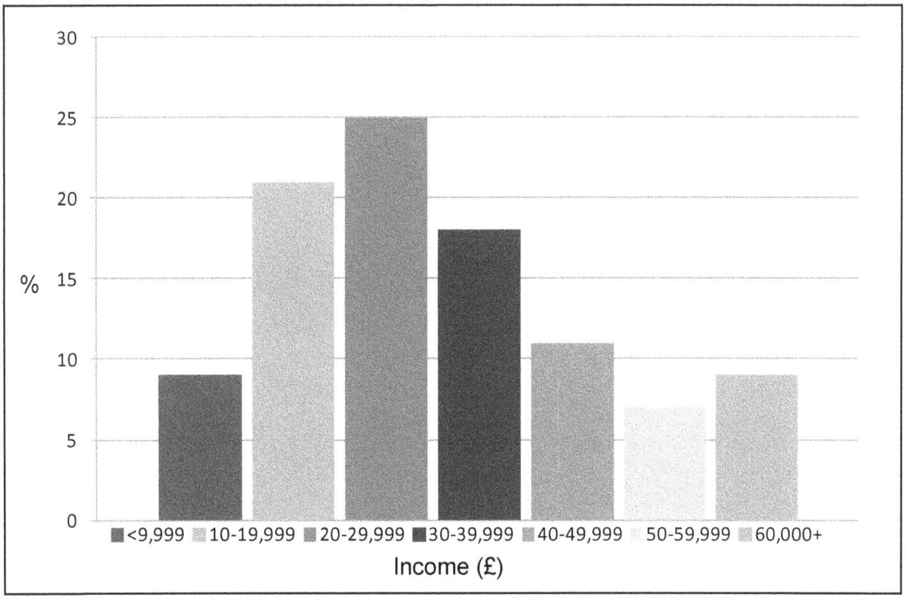

Figure 7.4: Household income of attenders at the Sidmouth International Festival

This chart shows the distribution of income of attendees at a music festival in the early 2000s. Each category has a width of £10,000, except the final one which is '£60,000 and above'. Note that the figures represent household income here, not individual income, as this was the nature of the question on the questionnaire from which the data and question have been derived. In relation to the column which is '£60,000+', it is important to note that although as many as 8% of the attendees have a household income of over £60,000, the information here does not tell us what proportion earns £60-70,000 and what proportion earns over £70,000, or even the proportion earning over £100,000. The category '£60,000+' was chosen, as it was assumed that a relatively small proportion of attendees would earn over £60, 000 and this is largely the case – it is just 8%. However, if it was decided that it was important to know the proportion of attendees earning, for example: £60,000-£70,000 or in excess of £100,000, more research with this group would be needed. Nevertheless, the responses to this question are a reminder that creating the best response categories possible in advance, is very important and that pre-testing and/or a piloting are key activities before employing the main survey.

However, it is relatively easy to see on this chart that the biggest single category is the income range of £20-29,999 with 25% of the sample in this category. It is also easy to see that a relatively small percentage of the sample had a household income under £10,000 and an even smaller percentage had a household income of between £50,000 and £60,000.

Pie charts can be used to show a variety of topics or themes. The one below (Figure 7.5) indicates the age range of attendees at the same music festival. Note that most of the categories are of equal size, except the first one which shows those between 18 and 20 years old. All categories from aged 21-60 are shown in ten-year age bands, whilst those aged over 61 are all shown together in the final category. What can be seen very easily in the chart is that more than half those attending the festival are aged between 41 and 60.

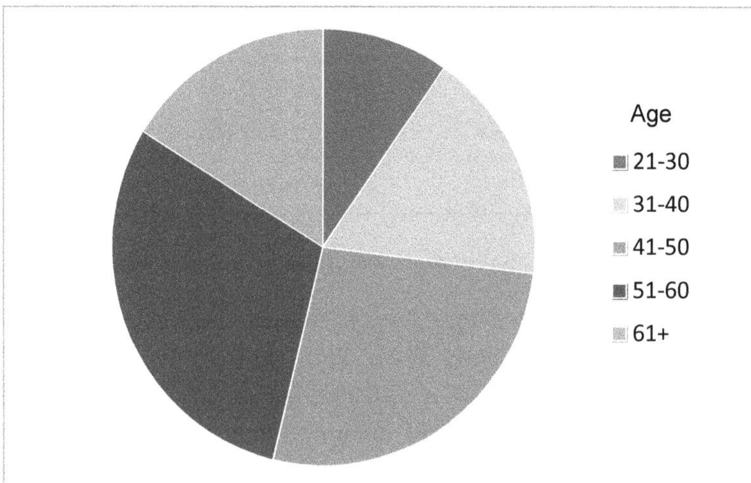

Figure 7.5: Age of attenders at the Sidmouth International Festival

It is also possible to combine different types of chart. So, below (Figure 7.6) is a pie chart combined with a bar chart. The chart indicates the origin of visitors to Stonehenge. The pie chart divides the visitors into two groups. The left-hand sector of the pie chart, representing 38.5% of visitors in the survey I conducted (see Mason and Kuo, 2007) shows UK visitors. The right-hand sector of the pie chart (61.5%) shows international visitors. The bar chart breaks down the international visitors into the four different groups represented in the survey, 'Americas', 'Other European',' Asia and Pacific' and 'Africa and the Middle East'. The percentages for each of these four regions is shown. Note that the bar chart adds up to 61.5% as this is the total percentage of 'international visitors'. The lines running from the top and bottom of the bar chart to the pie chart indicate that the bar chart is a breakdown of the right-hand side of the pie chart only. This chart combines a pie chart with a bar chart, but it is also possible to have a pie chart with another pie chart next to it which shows the same information as in the bar chart in Fig 7.6. The chart below in Figure 7.6 is known as a 'bar of pie' and one using a large pie chart and a smaller pie chart of a segment of the larger, is known as 'a pie of pie'.

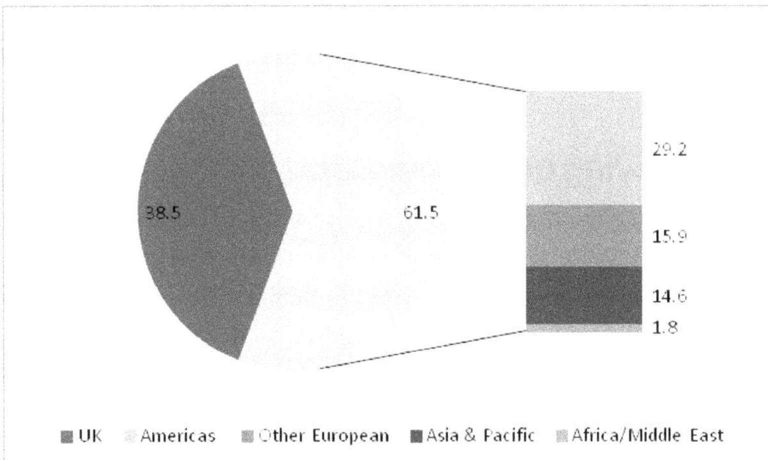

Figure 7. 6: Visitors to Stonehenge, normal place of residence

There are several other types of chart that can be used and also different types of graph than those presented above. Both PCs that use Microsoft, and Apple Mac systems have sophisticated software packages that enable pie charts, bar charts, histograms as well as different types of graphs to be created. These usually operate with Excel and are relatively straightforward to use.

7

Presenting qualitative data and results

Unlike quantitative research there is not a set range of ways, (which would usually include the use of bar charts, pie charts or statistical calculations), of displaying qualitative data. A key reason for this is that, by definition, your data is qualitative and is not generally amenable to numerical conversion.

It is the responses to open-ended questions and where you may have prompted respondents to comment more and/or explain their specific responses, or to add to something they had said previously, where the quantitative data display approach will be of no use, as the responses are completely in the words of your respondents.

Nevertheless, you should be aware that not all qualitative researchers follow the trend of refusing to use what most would regard as a quantitative presentation technique. Hence, there is not usually a problem with putting at least some of your data in a table. For example, you may want to show some demographic information regarding your interviewees or focus group attendees and using a table would be a good way of presenting this. Some of your questions may have been, at least, in part, closed-ended, such as those where you may have asked a factual question with a follow up explanatory part. It may be possible, and even sensible, to put the closed-ended part of the interview results into a table.

■ Displaying qualitative data

Despite qualitative data not requiring particular types of graphical presentation, there are a number of conventions that most qualitative researchers use when initially presenting their data. This initial presentation, or data display, will involve some form of analysis, although it may appear in a chapter in the dissertation titled 'Results' – implying no analysis has taken place.

However, before proceeding with presentation, prior to analysis, you will need to prepare your data. If you have conducted interviews or used focus groups and used some form of recording device, you will need to decide what to do with the recording. The great majority of those involved in qualitative research use some form of electronic recording device and then transcribe the spoken word, into written words. What you need to do, if you have recorded respondents' answers, is decide if you will do this. If not, you will have to analyse from recorded spoken material. There is a major advantage of this, which is you have not made any changes to what your respondents have said – the data is 'pure and uncorrupted'. However, it is an extremely difficult and time-consuming task, and will almost certainly require you to listen again (and again!) to the recording.

But if you decide to transcribe, and most researchers do, how exactly should you do this? As Long (2007) indicates there is a number of approaches that can be used. The most sensible way, you may consider, is a word-for-word transcription from spoken to written forms. However, this can be time-consuming as noted earlier in the book, with researchers generally agreeing that, for example, one hour of recorded interview can take anywhere from three to five hours to transcribe (see Long, 2007: Veal, 2011). An alternative approach, avoiding a word-for-word transcription, would be to type up a summary of what your respondents have said. This will almost certainly save transcription time but has a key disadvantage of losing the exact words respondents have used and also the way (such as the tone of voice) in which comments have been made. Respondents can, for example, strongly emphasise some comments and a summary may omit this. However, yet another way to transcribe is word-for-word, but in addition the researcher adds in comments such as 'long pause before speaking', or 'respondent looks embarrassed', or 'respondent has angry expression'. You could, of course, give your recordings to someone else to transcribe – this will certainly save you time, but will almost inevitably cost you money. There are some disadvantages to getting someone else to transcribe your recording – for example, they will not necessarily pick up on nuances in responses. Of greater significance is that if you do the transcription yourself, however slowly you type, you will become familiar with responses and this familiarity is an essential aspect of the analysis process that will follow the transcription phase.

Familiarising yourself with your data is quite likely to require more than one, or in fact, several readings of this material. It may be only after, for example, three readings that you are beginning to get some idea of terms that occur frequently, and the nature of certain responses and respondents in relation to other responses and respondents.

Summarising

Although you may include transcripts of your interviews, or observations, this information alone will not lead to positive assessment from your supervisor or examiner. You need to present at least some form of summary of your qualitative results. This means that you will have to carefully read your transcripts of what your respondents have given you as a response to each of your questions. You may then want to compare all the answers that one respondent has given to all your questions and possibly look for certain aspects of their responses to help you put this specific respondent in a specific grouping with other similar respondents. You will also need to compare answers from different respondents to individual questions. At this point your analysis is not attempting to explain why you have received the answers, but only attempting to group and summarise, with the main purpose of making your writing easier to read and comprehend.

The most common way to present qualitative data, via a process of summarising the answers you have been given, is to paraphrase responses. This is putting into your own words what respondents have said. It is quite likely that even with, for example, a small group of respondents in an interview or focus group context, you have received a range of views and comments. It is possible that you have a 'majority' view with say 14 out of 20 respondents giving this and a minority view with, let's say, four respondents providing this view and two others who do not appear to have strong views either way. You should summarise both the majority and minority views. However, you should also use selected quotations to back up the summary points. These quotations, the verbatim statements and, as indicated above sometimes termed just 'verbatims', should be used to illustrate the actual words of respondents in your summary comments. In this way they should be typical comments of the particular summary point you are indicating. You should, where you believe it appropriate, also present some atypical or unusual direct quotations. One important reason for this is that in your summarising you may tend to ignore, or at least play down, the unusual and atypical comments. Your summaries may appear to be presenting too much of your view of the results and not the data themselves. In other words, you may appear to have over-interpreted in your summarising.

What the previous paragraph has indicated is that you should be aware that summarising views is not necessarily an easy process, as you will be attempting to be as true as possible to the original statements of your respondents, but also summarise a range of these views. Inevitably in the process you will 'smooth off the rough edges' of individual comments to achieve the summary. On the other hand, if you do not summarise your write up, it may appear very disjointed with little clarity on the key points your respondents have actually provided and could give the impression that you do not understand what respondents have said.

Presenting results from mixed methods research

As mixed methods involves the use of both quantitative and qualitative techniques, presentation of results should follow the formats suggested above in each of the two sections. However, you will need to decide on the order of presentation. It would probably be best if presentation was in the chronological sequence of the use of a technique, be it quantitative or qualitative. Nevertheless, you may have good reasons for presenting in a different sequence. If so, you will need to provide the rationale for this.

Writing this chapter in your study

What you present as 'Results' in will clearly depend on whether you have conducted quantitative or qualitative research. A quantitative study is likely to lead to the use of tables, graphs and charts for the presentation of results. You will need to decide which particular graphical form of display fits your results best. When presenting you should ensure that your results are clearly displayed, but you should also make them as attractive as possible.

You will need to make decisions on the order in which you present the results. The most obvious way is to present it in the order that your questions appear on the questionnaire. However, it is not always the most appropriate way. There may be a strong argument for using a thematic approach, particularly if you have linked questions around specific topics. There may also be an argument for presenting all demographic material separately and perhaps having this displayed before other results, even if the questionnaire provided demographic questions after the research topic questions.

When creating diagrams, it is best not to number them initially. It is very likely that you will increase or decrease the number of such diagrams, so will have to change the numbering system. So, write something such as 'Figure X' until you have reached the final draft of your dissertation.

It is considered important by many researchers (see discussion of this in for example, Veal 2011; Long, 2007) that your results are presented in such a way that the reader is not being led by you, the author, to specific conclusions in advance of the comments about your results in your 'Analysis' chapter or section. In other words, the reader should be in a position to 'make their own mind up' about the meaning of your results, without any comment from you. However, depending on the nature of your research, it may not be that straightforward to separate the results from the discussion of the results. If you do combine the results with discussion of results in one chapter, you should explain carefully why you are doing this.

If you have used a qualitative approach, the way in which you display your results will depend very much on the data you have collected. Display of data may require little more than a summary of the key points made by respondents and the use of some 'verbatims' to illustrate this. However, some results in qualitative research, such as demographic data, can be displayed in tables, charts or graphs.

Depending on the nature of your qualitative research, even more so than in the case of quantitative research, it may not be that easy to separate the results from the discussion of the results. This will be particularly the case if you used, for example, emergent research (see Creswell, 2009). This form of qualitative research is where you modify your techniques and/or questions on the basis of results and findings from an earlier stage of research. If this is the case then it is very likely that you will need to discuss why you made the changes – this will inevitably involve discussion, and not just presentation of the results from the earlier stage. However, as with the statement above concerning quantitative research, if you do combine the results with discussion of results in one chapter, you should explain carefully why you are doing this.

If you have used a mixed methods approach, the major decision will be which of the two approaches you display results from first. Probably the logical way will be in the order that you conducted the research. This is particularly so if the first approach produced results which were used to set up the next stage which used a different research approach and techniques.

Student activities

1 What are the advantages of using some form of graphical representation (e.g. bar charts, pie charts line graphs) when presenting quantitative data?

2 What do you understand by the term 'mutually exhaustive categories'? How would you use concept this in quantitative data display?

3 What are the reasons for calculating the mean value in relation to statements used in Likert scale questions?

4 Although Likert scales can be used to show a range of views, categories are frequently aggregated into 'Agree' and 'Disagree'. What is the rationale for this and what are the advantages and disadvantages of doing so?

5 What are 'verbatim statements'? When would you use them?

8 Quantitative Data Analysis

Introduction

We have already discussed the first stage of analysis. This is connected to the organisation and presentation of your results. During this first stage, you will be gaining familiarity with your results, and you may have begun to notice patterns in your data. The answers to certain questions may have stood out from the overall responses. You may have also seen differences in the answers of some respondents to the responses of others. The analysis process will involve you in finding similarities, clearly identifying patterns, confirming differences and providing explanations for your results.

The analysis process

When you are analysing in your study, you are continuing to ask questions. In the preparation of, for example, your questionnaire, you will have considered the questions to which you need to find answers in relation to your aim(s) and objectives. At the early stages of the primary research in your study, you will have prepared the questions you included on your questionnaire and may have piloted these in an attempt to reveal if they are appropriate questions, that respondents understand and to which respondents can give you useful answers.

In the analysis of your results, you will continue to ask questions, as you now need to consider, not just what your respondents have provided as answers, but what these responses actual mean. In other words, you want to find answers to the specific questions you provided, but more importantly at this stage of the dissertation, why you have received the answers that you have. This is likely to be the most difficult part of the dissertation and, you may not be surprised to find that, it is very likely to be the one that your supervisor and examiner will study particularly closely and assess very carefully. The section: 'Writing this chapter in your study' discusses this in more detail.

In the conducting of your analysis, you are attempting to explain your results (this is mainly trying to answer the question why?) so that you can get a better understanding of them. However, in attempting to understand your results the key process in which you are involved is *comparison*. Analysis is largely a process of comparison, and in your dissertation will involve:

- comparison between the answers of one respondent to each of the questions on, for example, a questionnaire;
- comparison between answers given by different respondents to each of the specific questions on the questionnaire; and
- comparison between your primary research and literature, including the research already published on the topic you are researching.

Conducting your analysis is likely to be the most important process that is required in your dissertation. The reason for this is that, through the processes of comparison outlined above, you will be attempting to answer the question 'What do my results actually mean?'

However, the approach you take to analysing your data will depend very much on the type of data you have collected. Also remember, as stated in Chapter 3, that it is always a good idea to decide on how you will analyse your data before you actually start to collect it, although this is not always done by students.

Questionnaire analysis

Probably the most common technique used by very many students, requiring quantitative analysis, is a questionnaire survey. Certainly, if your survey is made up mostly or completely by closed-ended questions, then it is feasible to analyse it in a quantitative manner. The first stage of quantitative analysis involves converting your responses into some kind of numeric form. In its simplest form this can mean for example, actual 'real' totals of respondents in a specific response category. The variety of questions on a survey means that there are a number of ways of displaying the results. The presentation of data requires the processes of summarising and organising material. This process of displaying actual figures from a summary of responses, and also converting these to percentages, has been discussed in Chapter 7 in relation to how to present your results. Likewise, the use of various diagrams (bar charts, pie charts and graphs) to present your results has also been discussed. However, as a reminder, it is important to be aware that actual numbers of respondents and responses to specific answers can be shown, although these are frequently converted to percentages, so that it is easy to compare results for different but related questions on the survey. Using percentages can also make it possible to compare your results with results from published sources including those in your literature review. Closed-ended questions such as the 'Yes/No' type can be displayed as percentages, while more complex closed-ended questions can be displayed using bar charts and/or pie charts. Some data may be appropriate to show as line graphs, or scatter graphs.

However, the presentation and mere summarising of your results is not what is required in the analysis chapter. It may be the first stage of analysis, but there is much more to do in attempting to answer the vitally important questions: What do the results mean? The following section discusses how you can answer this.

What do your results mean?

Some of your questions may involve factual information, including demographic-type questions where you will have asked for information on, for example, gender, age, income levels, education and place of residence. This will certainly be the case if you have attempted to achieve a representative sample and may have used some of these demographic factors to achieve this sample. Note that some of the questions you used to gather such information may have been in the form of checklists or 'tick box' type questions, but the common factor is they were intended to gather factual information and were closed-ended. Nevertheless, whatever the format of the closed-ended question, the comparison process here involves the comparison of answers from certain types of respondents to those of others and you may notice differences between some respondents' answers. For example:

- men may give different answers compared with women;
- older respondents may give different responses compared with younger ones;
- those with high education qualifications may give different responses to those with lower qualifications;
- some nationalities may give different responses compared to other nationalities.

It is also possible that respondents from the different groups indicated above may provide very similar responses, but you will not know until you study the responses carefully. Nevertheless, it is important to be aware that the process of comparison involves looking for similarities, as well as differences, and these similarities can be just as important as differences.

Factual type questions

However, as suggested above, the main 'factual type' questions on your survey may produce different answers from different types of respondents. I have used, earlier in the book, the example of my research at the Sidmouth International Festival. If I return to this now, one topic I was interested was the characteristics of those who purchased season tickets and day tickets. I wanted to find out if those who had purchased a season ticket (which included free accommodation at the Festival campsite) stayed in different accommodation (i.e. the campsite) to those who purchased a day ticket, which did not include any accommodation. So, in this case, I was comparing the type of ticket held with the type of accommodation of respondents.

I was also interested, in this particular research, in other factual responses, such as the method of transport to get to the Festival. Additionally, I wanted to discover how much money attendees at the Festival spent on food, drink and entertainment. The results from the responses to these questions were displayed in various diagrammatic representations. In terms of the analysis of them, I compared responses, for example, on the type of ticket and type of accommodation, and this revealed that a great majority of those who bought season tickets did stay at the Festival campsite, while the great majority of those buying a day ticket did not stay at the Festival site.

What I was also able to do with these responses to questions on ticket type, transport and how much money was spent on food and drink and entertainment, was to compare these 'topic' related question responses, with the demographic information I had collected. Hence, if I wished, I could focus in on very specific groups, such as 'males, between the ages of 51 and 60, with the educational qualification of a post-graduate qualification, with a household annual income of between '£50,000 – £60,000' to investigate if they had bought a season ticket or a day ticket. For this same group, I could also find out how much they had spent on, for example, food and drink or entertainment. There were many other groups (such as 'women, aged 31-40, with at least an undergraduate qualification and an annual household income of £21,000-30,000') with which I could compare the responses of with the group of males referred to above. There were, of course, yet other groups that could be compared with each other. This comparison of results, using demographic information and my main themes, was a major part of how I analysed my questionnaire responses.

When I analysed the data in this way, I could make some claims about my respondents and why they did what they had informed me about on their questionnaire. For example, in attempting to explain why the great majority of those with a season ticket stayed at the Festival campsite, I argued that their ticket type included free accommodation, so they would have had to pay (in effect twice) if they wanted different accommodation. So, my explanation was based on the evidence I had collected from respondents. However, in my attempt to use this evidence, I could not be entirely certain that this was necessarily, or indeed always, the explanation for their choice of accommodation. Nevertheless, I could use responses from other questions on my questionnaire, from those respondents who had season tickets to provide more evidence for their choice of accommodation. I could also compare my results with published literature, of a similar type involving festival attendees and season tickets, in an attempt to gain more evidence.

Note also that although I focused on identifying different groups and then comparing the results from these groups, as I worked my way through each respondent's answers to my questionnaire, I was building up a profile of individual respondents, and you will be doing the same thing in your study. This is particularly the case if you are carrying out a manual analysis of your results. If you are entering your results into a computer software package, such as SPSS, you may not be so aware of an individual's profile, but it is very likely that the computer programme will be able to identify specific characteristics and aspects of individuals.

Following on from the comparison of the results from different groups, a major part of my analysis of the study was to compare my results with other similar studies from literature. One such festival was an Australian music festival (the Woodford Music Festival – see Raybould et al., 1999) with a focus on folk music – the same focus as the Sidmouth International Festival. In fact, some of the questions I had used were based on the Woodford Festival, so it was possible to make a direct comparison between the results of the questions in the Sidmouth International Festival study and the Woodford Festival question responses. I also compared the results from the Sidmouth International Festival with literature concerned with other similar festival and event research.

So in summary this process of comparison involved:

- Comparison of the results to all of the different questions on the questionnaire of one respondent – this helps build a profile of individual respondents *
- Comparison of the results of one specific question on the questionnaire of different respondents, eventually involving all questions to

be compared – this helps to identify groups of respondents *

- Identification of different groups, according to demographic and thematic questions, and comparison of these groups in relation to responses to specific questions
- Comparison of the primary research study with published literature of a similar type.

* I regard the combination of these two analytical processes as analogous to a piece of woven cloth. The vertical strands of material in the cloth are specific individuals in your sample, whilst the horizontal strands are each individual respondent's answer to a specific question. Combined together, these strands are woven into a complex whole that is the key material for the analysis of your primary data.

In summary, in the above discussion, I have tried to make it clear how I used the evidence from my questionnaire to explain why respondents gave me the responses that they did. As stated at the beginning of the chapter, this is the major focus of this analysis chapter. So please be aware, that I was using evidence from data that I had collected, and I tried to link the specific topic questions I was focusing on, to other questions on my questionnaire to support my understanding of the evidence, and also looked at published literature in an attempt to find similar results. What is very important to be aware of is that none of this process involved 'guess work'. Also be aware, that I did not make a claim that: 'everyone who has purchased a season ticket will stay at the festival campsite'. Instead, I indicated, based on my evidence, that there was a *strong likelihood* that those who purchased a season ticket would stay at the Festival campsite.

Attitude-based questions

This discussion so far has focused on the analysis of closed-ended questions which are largely factual, be they demographic or thematic. However, it is very likely that you will have also used a number of attitude-based questions, which have involved an attitudinal scale such as a Likert scale or a bi-polar semantic differential scale. How can responses to such questions be analysed? In Chapter 7, we have seen how Likert Scale questions can be converted to numeric values. It is then possible to do relatively simple statistical calculations such as finding the mean for a specific statement. A similar process can be used in relation to other attitude scales and also bi-polar semantic differential statements.

Comparing the mean scores for different statements on a Likert scale is part of the initial process of analysis. The process of comparison, described above in relation to closed-ended questions, can also be used in relation to responses to attitudinal scales. So, this would involve comparing the results of one respondent to different attitudinal-scale questions on their individual questionnaire, which should help build up a profile of an individual respondent. Also, comparing the entire sample's responses to specific attitudinal questions would be part of this analysis process. Comparing the results from the primary research with similar studies from literature would, in addition, be part of the process.

Partly because attitudinal scales are concerned, not with factual information, but with what respondents believe, there are some limitations. In relation to the use of Likert scales, an important issue is that what is known as 'respondent acquiescence' (see Ryan, 1995). In the conduct of research using questionnaire surveys of different kinds (e.g. postal face-to-face, on-line), there is much evidence that respondents will tend to agree with statements, rather than disagree. This would seem to be very much linked to the psychology of respondents – it would appear that the majority of people are happier saying 'Yes' rather than 'No', particularly if they are completing the survey on a face-to-face basis. However, research has been conducted that shows personality type affects how individuals respond (see Berg, 1967 and Couch and Heniston, 1960). Such research suggests that there are certain personality types that will always prefer to answer 'Yes' and other types that will always answer 'No' to whatever the question! A way of trying to overcome this, is to use some 'reverse' questions. These are negative, rather than positive statements, such as: 'I do not like physical activities' rather than 'I like physical activities' (see Ryan, 1995). However, there can be a problem with this approach, as it is not always possible to create an opposite statement easily, such as by inserting a 'not' (Ryan, 1995).

Another problem relates to the middle of the scale. Strictly speaking, the middle point means 'Neither Agree, nor Disagree'. The 'Neither Agree, nor Disagree' position is intended to provide an option in which respondents do not have a strong view (either way) on a particular statement. It is debatable whether the mid-point of the scale should represent a negative position, particularly when it achieves score of 3 when converted to the 5-point scale (see Ryan, 1995). Also, many students (and indeed some respondents) view this mid-point as 'I don't know'. As indicated in Chapter 4, this is not quite the same as 'Neither Agree, nor Disagree'. 'I don't know' could imply the respondent has given no thought to the questions, while 'Neither Agree, nor Disagree', implies they have thought about it, but as stated above, they do not have strong views.

The number of response categories is another issue with attitudinal scales. From 'Strongly Disagree' to 'Strongly Agree', in the examples above, there are clearly five categories, (hence the scale numerically goes from 1-5). Some researchers (see Moser and Kalton, 1989) have indicated that this range of five categories may be insufficient. A scale with seven categories is therefore sometimes used. The key differences between this and the five-point scale is that there are only two sets of words used. At one end is 'Strongly Disagree' and at the other, 'Strongly Agree'. Some researchers (see e.g. Ryan, 1995 and Moser and Kalton, 1989) have suggested that however many categories (5, 7 or more), it may be better to use just numbers and these should just run from one end of the scale (indicated by 'Strongly Disagree') to the other (indicated as 'Strongly Agree') with no words between.

Yet another problem, which some researchers (see Clarke et al., 1998; Riley, 1996) view as a particularly important one, when using a Likert scale, is that words are being converted to numbers. Therefore, this is an attempt to convert the subjective to the objective. It is very easy to forget this when you are in the middle, for example, of a detailed comparison of the means for a number of statements on your Likert scale questions. If you have a mean of, let's say, 4.35 for the responses to one statement on a five-point scale, then this lies between 'Strongly Agree' and 'Agree' and most researchers will be happy to accept that it is a mean value for a statement that respondents do agree with. However, if you have a mean of 3.5, this is exactly half-way between 'Neither Agree, nor Disagree' and 'Agree'. So, what is the meaning of this in words? Clearly it is not easy to answer this question. With a mean of 2.5 (half-way between 'Disagree' and 'Neither Agree nor Disagree'), the same point can be made – it is not easy for this value to put into words. This perceived problem of converting words to numbers is an important rationale for those who use only numbers on a continuous scale between 'Strongly Disagree' and 'Strongly Agree', be it a five-point, seven-point or even nine-point scale. This problem also provides a rationale for the use of bi-polar semantic scales, which only use words at the extreme ends of an adjectival scale (such as 'dirty … clean', or 'wet … dry 'or 'hot … cold'), with no other words on the scale, but only numbers between the adjectives at each end.

Although it is customary to calculate the mean for Likert-scale questions, it is also possible and relatively common to provide percentages in relation to response categories. Table 8.1 indicates the major motivations for visiting the World Heritage Site of Stonehenge. Visitors at Stonehenge had a number of statements that they had to respond to on a five-point Likert scale from

'Strongly Disagree' to' Strongly Agree' where 'Strongly Disagree' was equal to 1 and 'Strongly Agree' equal to 5. Table 8.1 shows motivational factors that achieved a high mean and also those with a low mean value. This table also shows aggregated 'Agree' percentages (calculated by adding together the 'Agree' and 'Strongly Agree' percentages and aggregated 'Disagree' percentages (calculated by adding together the 'Disagree' and 'Strongly Disagree' percentages). The use of both a mean score and an 'aggregated' percentage in one table shows the link between these two and can also indicate, reasonably clearly, strength of view in relation to a statement. Hence, a high mean value is usually linked to a high percentage in the aggregated 'Agree' category and a low mean value is usually linked to a high percentage in the aggregated 'Disagree' category.

Table 8.1: Motivational factors for visiting Stonehenge

Motivational factor statement	Agree % [1]	Disagree % [2]	Mean
Stonehenge is unique	87	3	4.5
To expand my knowledge	75	8	4.1
I am interested in prehistoric monuments	76	7	4.0
Stonehenge is a World Heritage Site	62	17	3.7
It was recommended as a place worth visiting	61	18	3.7
I am interested in archaeology	55	22	3.6
Stonehenge is a fun place to visit	48	22	3.4
It is part of a tour of major UK tourist sites	47	33	3.2
To be with my partner/family	46	34	3.2
A day trip from home	47	33	3.1
Stonehenge has a spiritual value	37	39	2.9
To be with friends	39	40	2.9
To escape routine	39	46	2.8
Stonehenge has a religious meaning	43	46	2.5
I was just passing and decided to visit	19	69	2.0

[1] Calculated by aggregating responses in the 'Agree' and 'Strongly Agree' categories.

[2] Calculated by aggregating responses in the 'Disagree' and 'Strongly Disagree' categories.

Open-ended questions require a rather different analysis process, largely because there are no set response categories. Hence, responses to open-ended questions usually are in the words of individual respondents. Therefore, such questions should provide responses that are very similar to those given in an interview or focus group context, both of which involve qualitative research techniques. If you have used open-ended questions in your questionnaire survey, an appropriate way to analyse them is to be found in Chapter 9, which focuses on analysing qualitative data.

Statistical tests

This is not a text specifically on statistical techniques, so it does not go into detail on the great variety of tests that can be carried out. Also, there are a number of very good computer statistical packages for analysing quantitative data of which 'SPSS' is probably best known This is an abbreviation for Statistical Package in the Social Sciences and this package is more than adequate for most dissertations in tourism and related fields. However, if you are conducting quantitative research, being aware of statistical techniques and representations (whether computer-based or not) is important, as statistical tests can not only greatly enhance the reliability of your data, but also make you feel more confident in what your research has revealed. On the other hand, it may also stop you from over-claiming about certain results. Your supervisor and examiner are unlikely to respond positively if you make a claim that is not backed up by your data, or if you have failed to conduct a statistical test that would have enabled you to realise the significance and importance of your results!

Nevertheless, an important point to be aware of is that it is not compulsory to use statistical tests in relation to quantitative research. Even if you have decided in advance that you will be using statistical techniques, you need, and must provide, a rationale for this. An important part of this rationale is likely to be that when you have read through your results, you have seen connections or relationships or differences between variables. However, in this context, you are not entirely certain about the existence or nature of a relationship or differences and, therefore, the statistical test is to confirm your hunch (or possibly reject it).

As stated above, there are a great variety of statistics that can be generated from your data, and several statistical tests that can be conducted. But what is important and what is worth doing?

The first point to be aware if is what statistics can actually show. Three key aspects that statistics can reveal are:

- patterns in the data
- relationships or associations between variables
- differences between data sets and variables.

It may be the case that as you have glanced through your questionnaire results, you have noticed certain similarities in answers from specific type of respondents. So, for example you may have noticed that respondents in a particular age group provide responses which are similar to others in the same age group, but this age group's responses are very different from other age group's answers. Or you may have noticed that some of the female respondents have given different answers to the male respondents. Statistics will enable you to gain a very good indication of whether your 'hunches' are, in effect, correct and to what extent they are correct. Statistical tests are also very important as they can tell you whether the results you have obtained have occurred by chance or, for example, that there really are differences between your groups and each group is distinct.

In much large-scale quantitative research, because of the size of the sample, or because there are certain characteristics of the population which are known to occur in the sample being used, certain types of statistical test can be used, which may not be available to you in your project research. The type of test that can be conducted with a very large sample is known as a *parametric test*. Such a test, by definition, can only be used when the 'parameters' (e.g. certain demographic factors or characteristics of particular relevance to the research focus, such as 'regular diners at a restaurant on a specific day of the week', or 'certain types of spectators at a specific sports event'), of the sample are known and can be assumed to be the same as those of the entire population from which the sample has been obtained.

If your research involves a questionnaire with just sixty respondents queuing to watch an outdoor theatre event in Cornwall, or one hundred spectators of a major sporting tournament that attracts hundreds of thousands, or fifty diners from the five thousand who annually dine at a five-star restaurant each year, can you be sure your sample is representative of the entire population in each of these cases? It is highly probable that it will be difficult to respond to this question, but you should assume the answer is 'No'. The reason is that you are very unlikely to know the characteristics of the entire population for your specific research (if it is similar to examples above), so will not be able to tell if the sample is representative. This means that unless you are fully confident that you know the characteristics of your population, you should not make use of parametric tests. Instead, you should confine yourself to non-parametric tests. There are a number of these, but one is particularly useful in tourism related research and is discussed later in this chapter.

For many students, relatively simple statistics may be adequate in their study. These statistics may relate to just one variable – this could be, for example, a demographic factor, such as gender or age. However, it could also relate to the substantive topic of your research, such as whether an attraction has repeat visitors, or perhaps 'satisfaction with service quality', or possibly 'length of stay in a hotel', or may be concerned with attendance at a leisure centre. Such research is focusing on single variables. But relatively simple statistical analysis can be conducted into several variables to investigate if they are linked. It is quite likely that you will want to compare demographic responses with specific topics on the main part of your questionnaire. Hence, to take just two variables, you could look at whether there are differences in views on a specific topic between male and female respondents. This focus on more than one variable could also involve consideration of relationships in responses to specific substantive topics on the questionnaire. For example, you could be investigating whether repeat visitor have different views from those who are visiting for the first time. It is also possible, and frequently desirable, to investigate several variables at the same time. So, for example, you may want to investigate whether male season ticket holders, aged between 31 and 40, who have previously attended the music festival, which is the context of your research, give higher satisfaction ratings with the festival, when compared with female day tickets holders, aged between 21 and 30, who have not attended before.

It may be the case that you are not investigating differences between groups in your sample, but want to know if two variables are associated. For example, you may be interested in the relationship between income and frequency of holidays taken in a year, or alternatively age and participation in sporting activities. The particular tests that can be conducted to investigate associations depend on the nature of the data. If you have interval data, you can use a correlation procedure, known as Pearson's Product Moment Correlation Co-efficient usually abbreviated to just '*r*'. The statistics package SPSS and other computer packages can calculate *r*. The values of *r* should fall in a range between -1 and +1.

For ordinal data the most common test used is Spearman's Rank Order Correlation (summarised as '*rho*'). *Rho* could be calculated in relation to Likert scale attitudinal questions and other variables such as age, education or income. The values of *rho* should also fall in the range between -1 and +1. In both cases (*r* and *rho*) your result could be a positive correlation, where, as one variable increases, so does the other, or a negative correlation where it is the case that as one variable increases, the other decreases. In the case of a positive correlation the result should, be near to +1, with a negative correlation the result should be near -1. A result that is close to 0 means little, or no, correlation. It is very important to be aware, with both types of correlation, that there is not a cause-and-effect relationship – where one factor causes the other – just an association. When you are writing up your analysis you must make this clear and not fall into a trap of arguing that you have achieved confirmation of a causal relationship.

Descriptive statistics

There are certain key descriptive statistics that are almost always worth investigating, even if you plan to go no further than this. They will usually provide you with information that you can comment on which will add a dimension to your analysis. These descriptive statistics are:

- Frequency
- Average
- Spread
- Shape.

Frequency

Frequency relates to the number of responses in a particular category. This can be indicated simply by giving the number in a particular category. So, if you had 120 respondents in your survey with 66 females and 54 males you could simply put this in your 'Results' section. However, it is more common to find these figures as percentages, which are in fact relative frequencies. The percentages figures here, (with 66 females and 54 males) are females 55% and males 45%. However, it is always important to indicate what percentages are percentages of, and the following example should help explain this.

Using the information in the previous paragraph, (the example with 120 respondents) and assuming your study focus is on the age of workers in hotels, then let us assume that you have also gathered the following data:

- for staff aged under 30, the total is 72,
- for those aged between 31 and 50 the total is 30, and
- for those over 50, the total is 18.

The total sample of 120 is made up of:

- 38 female and 34 male aged under 30,
- 18 female and 12 males aged 31 to 50, and
- 10 female and 8 males over aged 50.

But what are the percentages?

The total percentages in each category are as follows:

- those aged under 30 make up 60% of the total sample,
- those between 31 and 50 comprise 25% of the sample
- those over 50, comprise 15% of the sample.

By converting your actual totals to percentages, you are able to indicate percentages of males and females in each category as well as a percentage of the entire sample. This shows us that:

- females under aged 30 comprise 52.7% of their age range, but 31.6% of the whole sample,
- females between 31 and 50 comprise 60% of their age range, but 15% of the whole sample,
- while females over 51 one comprise 55.5% of their age range, but 8.3% of the whole sample.

The figures for the males are as follows:

- they make up 47.3% of those aged under 30, which is 28.3% of the whole sample,
- they make up 40% of those aged 31 to 50, which is 10% of the whole sample
- they make up 44.5% of those over 51, which is 6.6% of the whole sample.

This example should indicate why converting to percentages is usually preferable to leaving results in your actual figures. However, there is also an issue when your sample size is less than 100. For example, converting 43 questionnaire responses to percentages means that if you use certain figures this can be misleading. So, for example, 2% of your sample is less than one respondent ('How can this be possible?' your examiner could ask), 10% is only just over 4 respondents and 50% is only 21.5 respondents!

Average

Finding the average is important in statistics as it indicates where the central point or the 'centre of gravity' (Long, 2007) of your responses is. The average, statistically, speaking, is the arithmetic mean or commonly referred to as just the 'mean'. So, knowing the location of the centre of gravity, or mean, as we have seen in several earlier discussions is very useful when using measuring scales which have two opposite ends, such as Likert scales or bi-polar semantic differential scales. Knowing the centre of gravity will enable you to recognise easily whether this is near one adjective on your bi-polar scale, or closer to the opposite adjective and if using a numbered scale, will provide you with a precise numeric value for this. As we have noted earlier, calculating the mean can be particularly important in relation to Likert scale questions, as once you have the mean value for a particular statement on your questionnaire, then it is easy to compare this with mean values for other statements. This will also enable you to compare results from one respondent to different statements in relation to the entire sample's mean for the statement. If you have replicated a study from literature, despite the fact that your study will have been at a different time, probably in a different location and almost certainly with a different sample, you will still be able to compare your results with this and any other similar published study.

Spread

The mean is also important because it is related to the spread of the data. Once a mean value has been calculated it is possible, as indicated above, to see which values are below the mean and which above this. The mean, in effect, divides the sample into two. Between the lowest value and highest value is known as the range. However, of particular importance, assuming there is what is known as a normal distribution curve, statistical probability theory indicates that if you were to draw a graph of all your values, with the mean at the centre of the graph, the data will be spread in a particular way. The shape of the graph is known as a bell-shaped curve – this can be seen in Figure 8.1. If the line which has the mean at the centre is marked out in regular distances from the mean, then, with normally distributed data, at what is known as one 'standard deviation' from the mean, just over two thirds (68%) of all values will be found. The term standard deviation is explained below. The concept of distance from the mean is important in terms of the standard deviation and usually requires reference to going 'above' the mean and 'below' the mean in numeric terms on the graph, where below the mean is to the left and above the mean is to the right. If we then move one more standard deviation on each side of the mean (so this is known as two standard deviations) then 95% of all values will be found and going one more standard deviation (this is now three standard deviations) from the mean in each direction – above and below the mean – then 99% of all values will be found. If this explanation is not fully clear, then the diagram (Figure 8.1) should help.

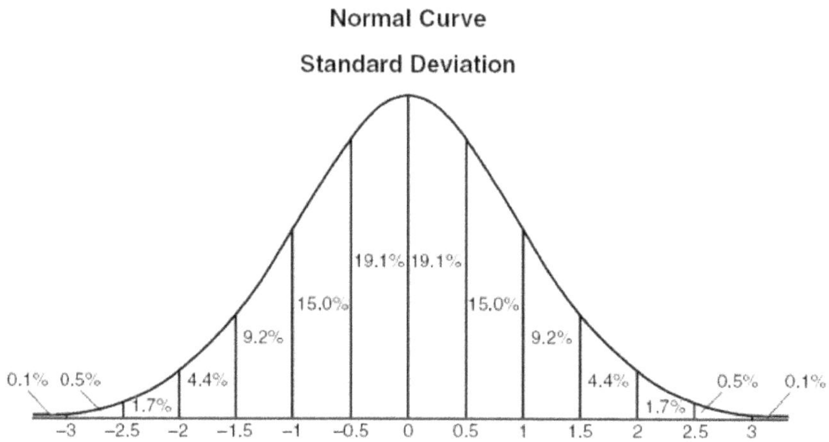

Figure 8.1: A graph of the normal distribution of data. This bell shaped curve also indicates the standard deviation from the central 'mean' position.

The term standard deviation has been used above and this is a very important concept in terms of statistical analysis. Standard deviation is a measure of dispersion. Measures of dispersion describe how scattered (or dispersed) data are. Conventionally, the data are subdivided into what are known as *quartiles* and these are represented on either side of the mean. In other words, the data are subdivided into four with two equal size quarters (known as quartiles) on each side of the mean. So, in effect, to the left side of the mean is divided into two equal areas and to the right side of the mean is subdivided, in the same way, into equal size areas. The data are then considered and looked at in relation to each of the quartiles, to see where the data are located.

If quartiles show how spread or dispersed the data are, there are two important measures of how the data is grouped around the mean. The two most important are *variance* and, the term we have just encountered, standard deviation. These are closely related terms. Variance is a mathematical concept which is calculated for any set of data in the following way. First of all, what is known as the mean deviation is calculated for all data in the data set. The mean deviation indicates how much, on average, the observations in a data set differ from the arithmetic mean. Putting this another way, the mean deviation tells us the average distance by which all items in a data set differ from their mean. If we take five numbers 3, 4, 5, 6 and 7, the range of these values (the difference between the highest and lowest) is 4 (from 3-7). The mean is 5 (this is calculated by adding 3+4+5+6+7 together and then dividing the total (25) by 5, which is the number of pieces of data in the set.

To find the average distance of all items in the set from the mean (the mean deviation) we take the numbers 3 and 7 and note these are both 2 points from the mean, whilst the numbers 4 and 6 are both 1 point from the mean. The remaining number is the mean itself (5). So, to calculate the mean deviation we add together the distances from the mean, which are as follows: 2+1+0+1+2. (Note here that we do not need to write on one side of the mean, minus numbers and the other side of the mean plus numbers). The total is 6 and there are five values in the data set, so the mean deviation is 6/5 = 1.2. Be aware that it is most likely that your data will be more complex than this small-scale example, but the principle of calculation is the same.

Having arrived at the mean deviation the variance and standard deviation are calculated in the following way. The variance is calculated by squaring all the deviations, in the data set, from the mean. These are then added together to give the sum of the squares (or the sum of the squared differences). The mean of this sum of the squares is known as the variance. In the case above, the sum of the squares is as follows: 4+1+0+1+4 = 10. (These figures are the result of squaring 2+1+0+1+2). The mean of the sum of squares is therefore 10/5 =2. The standard deviation is calculated by taking the square root of the variance. So, in the example being used, the standard deviation is the square root of 2 which = 1.41. It is worth noting that the variance is a value which is based on aggregating squared units, whilst the standard deviation (as it is calculated by taking a square root of the variance) brings the measurement back to 'ordinary' units, compatible with the original data. As indicated at the end of the previous paragraph, it is most likely that your data will be more complex than this small-scale example, but once again the principle of calculation is the same.

There are two other terms related to averages which are referred to in texts on statistics. These are the *mode* and the *median*. The term 'mode' refers to the most frequently occurring value in your data. This may be useful in your study. The median is the exact mid-point of a particular data and is not very likely to be useful in tourism related studies. However, in terms of the mode, because it is the most frequently occurring value, all other values will occur less frequently than it, and you can use this as a form of benchmark to compare specific values with the mode. In Likert scale statements, and when you have made use of bi-polar semantic differential response scales, the mode can be particularly useful as it will provide an 'easy-to-see' indication of where the most frequently occurring value for each type of scale is found. Additionally, once the mean for a Likert Scale statement, (or the mean in relation to the adjectives at opposite ends of a bi-polar scale), has been calculated, the mean and mode for each can be compared. It is frequently the case that the mean and mode for a specific statement will be similar, with the mode being a whole number and the mean likely to be a whole number, plus at least one decimal point (see the examples below). In terms of the median, being the midpoint of your particular data set, if it is seen as necessary, calculating this will enable you to discuss values below the median and above the median.

Below in Figure 8.2., there are a number of examples to show the relationship between mean and mode. The examples are all using Likert scale questions. In each of the examples, there is a sample of sixty respondents and the assumption is that in each case, they have responded to a statement using a Likert scale where '1' = Strongly Disagree, '2' = 'Disagree', '3' = 'Neither Disagree nor Agree', '4' = 'Agree' and '5' = Strongly Agree'. In Figure 8.2, the top row in each case is the Likert scale value and the second row, immediately beneath the first is the number of respondents in each category 1-5.

Figure 8.2: The relationship between mean and mode

Example 1

In this example, there are no responses in the categories 1 or 5, and the response totals for categories 2, 3 and 4 are shown below.

1	2	3	4	5
0	15	30	15	0

Calculating the mean is as follows:

1×0 = 0, 2×15 = 30, 3×30 = 90, 4×15 = 60   Grand Total 30 + 90 + 60 = 180.

As the total number of respondents = 60, the mean is: 180/60 = **3.**

The mode is also **3** (this is most frequently occurring value in the responses)

Example 2

The responses are as shown below

1	2	3	4	5
4	6	20	18	12

Calculating the mean is as follows:

1×4=4, 2×6=12, 3×20=60, 4×18=72, 5×12 =60   Grand Total 4+12+60+72+60= 208.

As the total number of respondents = 60, the mean is therefore 208/60 = **3.47.**

The mode is **3** (the most frequently occurring value).

Example 3

The responses are as shown below

1	2	3	4	5
5	5	12	24	16

Calculating the mean is as follows:

1×5=5, 2×5=10, 3×12=36, 4×24=96, 5×16=80 Grand Total 5+10+36+96+80 = 227.

As the total number of respondents = 60, the mean is therefore 227/60 = **3.8.**

The mode is **4** (most frequently occurring value).

Example 4

The responses are as shown below

1	2	3	4	5
24	18	10	6	2

Calculating the mean is as follows:

1×24=24, 2×18=36, 3×10=30, 4×6=24, 5×2=10 Grand Total 24+36+30+24+10 = 124.

The total number of respondents = 60. The mean is therefore 124/60 = **2.06.**

The mode is 1.

Note in each of the four examples above, that the mean and mode are usually fairly similar values, but this is not always the case as can be seen in the example below.

Example 5

The responses are as shown below

1	2	3	4	5
28	6	4	10	22

Calculating the mean is as follows:

1×28=28, 2×6=12, 3×4=12, 4×10=40, 5×22=110. Grand Total 28+12+12+40+110=202.

The total number of respondents = 60. The mean is therefore 202/60 = **3.36.**

The mode is **1**.

In Example 5, of Figure 8.2, the mean and mode are not similar; the mode is '1' while the mean is closest to the mid-point value of '3'. These values indicate that a large number of responses are not grouped around the mid-point but located close to each end of scale.

Shape

Shape is also an important factor when considering the distribution, or spread, of data. Graphical representations of how values of specific variables are distributed may reveal some very different shapes and patterns. Such graphs can reveal an even distribution, or peaks in specific places, or they can be skewed in a particular direction to give a lopsided shape. Looking back at Figure 8.1, a symmetrical distribution is shown. Here, the data is distributed evenly on either side of the mid-point of the central peak. In this case, the mid-point (median), the highest value (mode) and the mean all coincide.

However, the two graphs below (Figure 8.3 and Figure 8.4) show *skewness*, and also indicate what is known as *kurtosis*. Skew indicates how symmetrical the distribution is and kurtosis how peaked or flat it is. Figure 8:3 shows a positively skewed distribution. In this case the mean would be well to the right of the mode which is represented in this figure by the peak and the median would be somewhere between these two.

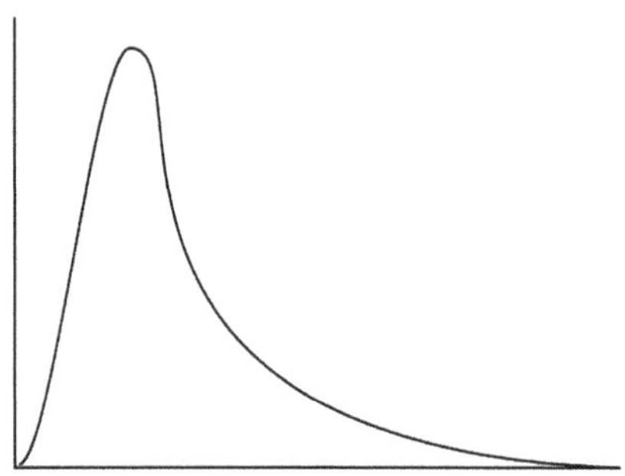

Figure 8.3: A Positively skewed distribution

Figure 8.4 shows a negatively skewed distribution. In this case the mean would be well to the left of the mode which is represented in this figure by the peak, and the median would be somewhere between these two.

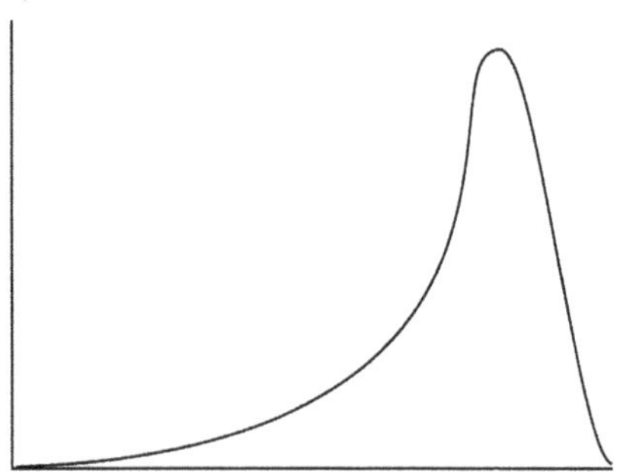

Figure 8.4: A negatively skewed distribution.

Figure 8:5 shows a bi-modal distribution, in which there is not one peak, but two. Here there are two equal modes, and the mean and median are located between the two peaks. Example 5 in Figure 8.2 would produce a distribution curve that resembles, to some extent, this bi-modal graph, although the peaks would not be of the same height as is shown in Fig 8.5.

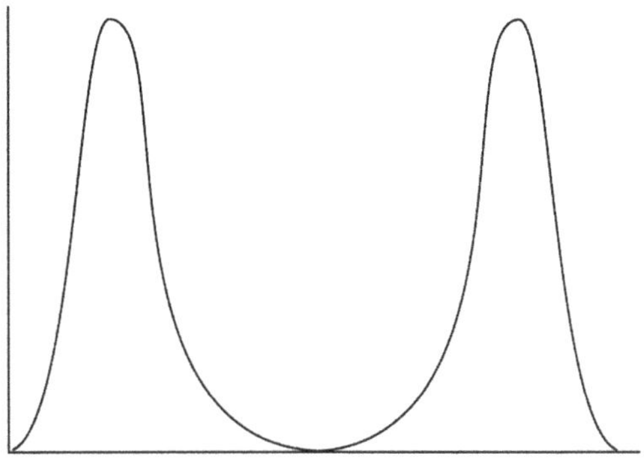

Figure 8.5: A bi-modal distribution.

The shape of distribution curves is particularly important when comparing the distribution for two or perhaps more groups. The shape of the curve provides an instant visual clue to the distribution, and it will be easy to see any differences in distribution.

The discussion, so far, has focused on various statistical measures, usually referred to as descriptive statistics, which are concerned with central tendency and dispersion of data, such as the mean and mode. However, there are a number of important statistical tests that can be carried out on quantitative data, some of which have been referred to above but not discussed in any detail.

Statistical tests

A major reason for conducting a statistical test is to reveal whether or not chance has contributed to results that have been obtained. Usually, the researcher will be hoping that chance can be eliminated and there is a statistically significant aspect to the collected data. A statistical test may be used to reveal differences between samples. Alternatively, a test can be used to reveal relationships between samples. Two commonly used tests are discussed in some detail in the next section – one is for use with parametric data the other with non-parametric data.

Student's t-test

Student's t-test[1] is a frequently used test in tourism and related fields and features in a good number of students' research projects. It is not a particularly difficult test to conduct. It is a test that should only be used on parametric data. If you are not certain that your data are parametric then it is not an appropriate test. If you know your data are parametric, then you can make use of this test.

Student's t-test is a test to find out if there is a significant difference between the means of two samples. Putting this another way, the t-test is attempting to reveal if two samples come from the same population. As stated above, an assumption has to be made that the data are normally distributed or are as near normally distributed. The data also have to be of the interval kind.

In the example discussed here, the t-test was used with responses from a questionnaire conducted at the World Heritage Site of Stonehenge (see Mason and Kuo, 2007). The main aim of the research was to get visitors' views on what Stonehenge offers to visitors and their experience there. A number of demographic questions was posed, as well as questions concerned with the main themes of the research. The t-test was used in an attempt to find any differences in responses to thematic questions, in terms of various demographic factors, but also variations in terms of different types of visitors and their views about thematic issues in the research. As can be seen in Figure 8.6, there are some differences between male and female visitor responses. Although these are not great differences, the t-test indicated that there is a statistically significant difference for the two question statements shown. The questionnaire collected information from both British and International visitors and some differences can be seen in Figure 8.6 and these are statistically significant differences. In fact, what the t-test revealed, and this was not obvious before the test was conducted, is that there was a specific group of respondents who are British, and they are repeat visitors. This group indicated that they have different views to other groups (such as international visitors and first-time visitors) and this was of particular importance for this research,. Why is this 'of particular importance'? It may be possible to suggest, solely from a research perspective, this is an interesting finding in its own right. However, I suggest it has important implications. This group are British and repeat visitors – they have somewhat more

1 The test was developed by William Gosset, a chemist working for the Guinness brewery in Dublin, and used to monitor the quality of stout. His pen name was Student.

negative views on the visitor experience at Stonehenge. So, it is possible to make a recommendation that attempts at managing Stonehenge should take note of this group and their views. It is not just managers who should be aware of this group, but those involved in the marketing of Stonehenge. A question that marketers may want to consider is, "does this specific group merit special attention in terms of marketing material directed at them?" Such recommendations, following on from the identification of the sub-group of visitors, would be ideally located in the final chapter – the Conclusions – of your study.

In an attempt to find any other differences in results from this questionnaire used at Stonehenge, a number of t-tests were conducted in relation to other responses. No other statistically significant differences were found. You may be thinking that perhaps there should be more statistically significant differences. In fact, it is very possibly the case that, if there are only relatively few differences in responses, this can be used to argue that your questions are 'good' (or more accurately, can be considered 'reliable'). If you had found many differences in responses, between different groups of respondents, this could suggest that some of your questions were not as good as they should be, and perhaps your whole questionnaire design was not clear enough and may not have produced reliable responses

Figure 8. 6: Student's t-test results for Stonehenge questionnaire survey

Likert Scale statement and t-test factor	Mean	Standard Deviation
Stonehenge is unique		
Female visitor	4.64	0.762
Male visitor	4.37	0.902
Not allowing visitors to touch the stones is necessary to conserve the site		
Female visitor	4.32	1.170
Male visitors	3.90	1.758
Current entrance chares are acceptable		
International visitors	3.53	1.298
British visitors	3.16	1.170
Entrance charges should be increased, with the extra money used to help conserve the site		
International visitors	2.78	2.37
British visitors	1.693	1.212
Access to Stonehenge should only be for archaeologists, historians or those with special permission, such as religious groups		
First time visitors	2.02	1.498
Repeat visitors	1.49	1.120

8

Current entrance charges are acceptable		
First time visitors	3.47	1.273
Repeat visitors	3.05	1.301
Visitors should be encouraged to go to a visitor centre rather than Stonehenge itself		
First-time visitors	2.27	1.543
Repeat visitors	1.65	0.991

Student's t-test is a parametric test. However, it is not always possible to confirm that your primary data are of the parametric type. Nevertheless, there are a number of non-parametric tests that can be used with non-parametric data and one that is used commonly in tourism and related studies is the Chi-squared test

The Chi-squared test

Earlier in this chapter there was discussion about the nature of the data you have collected and how representative your sample is of the population in your study. If you know your sample is truly representative, then you can feel confident in the use of a parametric test. If, however you have used a convenience sample, purposive sample or perhaps obtained your data by a 'snowball' approach, it is unlikely that you have a parametric sample. So, should you ignore statistical tests? The answer is a qualified 'No'. You should only conduct a statistical test if it is worthwhile – you may have noticed certain relationships or association in your data as you prepared them for presentation. For such data, there are a number of non-parametric tests that can be used. These do not assume a normal distribution in your data set, but they are still able to reveal statistically significant relationships or associations.

The Chi-squared test is one example of a test that can be used on data that does not have a normal distribution. Hence, it can be used on non-parametric data, and this can be nominal data. The Chi-squared test examines whether or not there is an association between one characteristic and another. In effect, Chi-squared is used to investigate if there is a relationship between one variable and another. However, it is important to be aware that even if the test reveals a statistically significant association or relationship, it does not mean that there is necessarily a causal relationship – in other words it does not mean that one variable has caused something to happen with the other. It means just that there is a relationship! A relatively simple example of the use of Chi-squared is shown below in Case Study 8.1.

Case Study 8.1: An example of the Chi-squared test

In this case study, visitors to a heritage attraction have been asked to indicate if they were satisfied with their visit. The table shows two types of visitor – first time visitors and repeat visitors. In this example, there is exactly the same size of sample – 100 in each category. This is to make the calculation easier, but different size samples can be used. Just looking at the results seems to show differences – the repeat visitors appear less satisfied than the first time visitors. The Chi-squared test can be used to indicate if there is a statistically significant difference.

Table 8.2: Types of visitors

Satisfied with Visit	Repeat Visitors	First time visitors	Total
Yes	45	60	105
No	55	40	95
Total	100	100	200

Table 8.2 is known as a contingency table and it is a 2 × 2 contingency table – there are two types of visitor and they were satisfied with their visit or not. As a 2 × 2 table there are four main cells of information: 'Repeat visitors, Yes', 'Repeat visitors, No', 'First-time visitors, Yes' and First-time visitors, No'.

In the discussion below of how to calculate frequency, the four main cells are as follows: 'Repeat visitors, Yes' is referred to as Cell A, 'Repeat Visitors, No', is referred to as Cell B, 'First-time visitors, Yes' is referred to as Cell C and 'First-time Visitors, No' is referred to as Cell D.

Prior to conducting the test which will involve calculation, we have to state a null hypothesis. As we have seen earlier, the null hypothesis is almost always a negative statement, so in this case it states that there are no differences between first time visitors and repeat visitors and their satisfaction with their visit. Or putting it slightly differently 'there is no association between the type of visitor and whether or not they were satisfied with their visit'.

The first step in the calculation is to consider two sets of numbers: one is what we have actually found in our study – known as the 'observed frequency' and the other is known as the 'expected frequency'. We have the actual figures for the 'observed frequency', but not for the 'expected frequency'. This expected frequency is what we would expect, based on the null hypothesis, in other words, if there were no differences in results. We need to calculate the expected frequency (and remember this is linked closely to the null hypothesis) – it is the result we would get if there were no differences between the two sets of data.

How to calculate the expected frequency is shown in Table 8.3. This table shows the four major cells (A, B, C and D) and in each of these cells it shows the process used to reach the specific number, which is the expected frequency. Table 8.3 shows the following: that N1 is the total number of 'Repeat Visitors', N2 is the total number of 'First-time visitors', T1 is the total number of those who have responded 'Yes' and T2 is the total number of those who have responded 'No'. N1, N2, T1 and T2 are important in the calculation for each of the four cells and also, so is T, which is the grand total of all those involved.

Table 8.3: Calculation process for expected frequency

Satisfied with Visit	Repeat Visitor	First-time Visitor	Total
Yes	Cell A = (N1 × T1) / T	Cell C (N2 × T1) /T	T1
No	Cell B = (N1 × T2) / T	Cell D (N2 × T2)T	T2
Total	N1	N2	T

The expected frequency of Cell A is (100 × 105)/ 200 = 52.5
The expected frequency of Cell B is (100 × 95) / 200 = 47.5
The expected frequency of Cell C is (100 × 105)/ 200 = 52.5
The expected frequency of Cell D is (100 × 95) / 200 = 47.5

These figures have been inserted into the cells and are shown in Table 8.4.

Table 8.4: Calculated totals for expected frequency

Satisfied with Visit	Repeat Visitor	First-time Visitor	Total
Yes	52.5	52.5	105
No	47.5	47.5	95
Total	100	100	200

The final stage of the calculation now involves comparing the observed frequency with the expected frequency. The observed frequency is our actual results (these are shown in Table 8.5). Table 8.5 indicates how we calculate the Chi-squared statistic. The left hand column in Table 7.5 shows each of the four cells. So in Cell A for the 'observed minus the expected' (O-E) it is 45 – 52.5 = -7.5 (note that this is a minus number, but will not be problem in the calculations, as is indicated below). For Cell B the figures are 55-47.5 = +7.5. Table 8.5 also shows the calculations for cells C and D. The third column shows the total for O-E in each cell which is then squared (this is O-E^2). In each of the four cells the O-E value is 7.5 (either -7.5 or +7.5). For O-E^2, the number 7.5 is multiplied by itself (this is squaring the number). Note when your square a minus number, in mathematics, the minus sign disappears and becomes a plus! So in each of the four cells $(O\text{-}E)^2$ = 56.25 (7.5 x 7.5). The very last part of the calculation is to divide the $(O\text{-}E)^2$ total for each cell by E, which is the expected frequency in relation to this cell. So for Cell A, it is 56.25 divided by 52.5, which equals 1.07 (to two decimal places). For Cell B, it is 56.25 divided by 47.5, which equals 1.18 (to two decimal places). Calculations for Cell C and Cell D are also shown in Table 8.5. Now it is just a matter of totalling the four numbers in the last column. It is 4.5 and this number is Chi-squared.

Table 8.5: Calculation of the Chi-squared Statistic

Cell	(O-E)	$(O-E)^2$	$(O-E)^2/E$
A	45-52.5 = -7.5	56.25	1.07
B	55-47.5 = +7.5	56.25	1.18
C	60-52.5 = +7.5	56.25	1.07
D	40-47.5 = -7.5	56.25	1.18

X2 (Chi-squared) is the total of the last column $(O-E)^2/E = 4.5$

So we now have a value for Chi-squared, but what does it mean? What we need to do first, before finding out what the value of 4.5 means, is to determine the relevant number of what is known as 'degrees of freedom'. This is a computed number that indicates aspects about the size of the data set and/or the number of values involved. The number of values in a table and the size of the table are important in terms of the calculation of degrees of freedom. The calculation is as follows:

> the number of categories for the row variable, minus one, multiplied by the number of categories for the column variable, minus one.

In this example, there are only two categories of row variable and two categories of column variable – the type of visitor (Repeat or First Time) and whether or not they were satisfied with their visit (Yes/No). To calculate the degrees of freedom is therefore as follows:

> row total minus 1 (= 2-1) multiplied by column total minus 1 (= 2-1), so $1 \times 1 = 1$

So the degrees of freedom in this case equals 1.

The very last thing that has to be done is to look at the applicable table for the Chi-squared (X^2) test. However, we need to decide at what confidence level we are prepared to accept the result. The confidence level is a measure of probability*. For any statistical test there is a published set of tables. These tables allow the researcher to decide whether or not the result they have obtained from a particular test is significant and at what level of confidence. This level of confidence, or significance level, is shown in the statistical tables as a probability value. It is usually shown as a decimal such as 0.05 or 0.01. The 0.05 level indicates a 95% confidence level and 0.01 a 99% confidence level. The 95% or 0.05 confidence level is particularly important as it is usually the lowest level, or minimum level, for deciding upon whether or not the null hypothesis should be accepted or rejected. The 95% level of confidence means that there is only a 1 in 20 possibility of the result occurring by chance. If we are prepared to accept our Chi-squared value from the example above at the 95% or 0.05 level, then if we look at a set of statistical tables it indicates that for Chi-squared to be significant at the 95% level its value must be equal to or greater than 3.84. The result we have obtained is 4.5 – this is greater than 3.84. Therefore we know we have a statistically significant result, which indicates that there is an association or relationship between the type of visitor and whether or not they are satisfied with their visit. As the result is significant, we can now, with confidence, reject the null hypothesis.

*The concept of confidence levels has been used for the first time here. In this case and, almost certainly in your own research, data comes from a sample, not the whole population (the relevant population from which your sample has been drawn, not the entire population of e.g. the UK). Confidence levels refer to how much confidence you can have in your sample being representative of the entire population. It is possible to use the sample you have obtained to calculate a range within which the values for the whole population are likely to be found. The values at each end of this range are the confidence levels. Although different degrees of confidence can be used, the most common is the 95% confidence level. It is within this level, that 95% of corresponding population values will be found (Long, 2007).

In summary, Case Study 8.1 shows how the Chi-squared test has been used to test what appears to be a relationship in the data in the study concerned with the satisfaction in their visit of two types of visitor. Until we conducted the Chi-squared test, there was not much more than a hunch that a relationship existed. Having conducted the test, it is possible to conclude that there is a statistically significant relationship. However, as stated earlier, it is important to remember that the test does not prove there is a causal relationship, (i.e. that one of the variable causes something to happen to the other), it only confirms a relationship or association. It is also important to be aware that the example used above is relatively simple and you are likely to have more complex data. However, if your data is more complex, it may be more difficult to initially see any relationships, but this could mean a greater rationale for the use of the Chi-squared test as it may confirm, or indeed, reject that there is a relationship.

Multivariate analysis

The discussion so far has focused on trying to reveal, via statistical analysis, if there is an association between certain variables, or if there are differences between them. This has concentrated on just two variables. However, it is possible to conduct other statistical tests that may reveal a more complex set of relationships and differences between respondents than has been previously discussed. When more than two variables are being compared, this is referred to as *multivariate analysis*. Using this type of analysis, for example, it is possible to select a number of demographic factors, such as gender, age, ethnicity and education, and consider these in relation to the major topics of your primary research. Multivariate analysis has a range of techniques, one of which is factor analysis. Factor analysis is designed for sets of means grouped by more than one classifying variable (Veal, 2011). Another type of multivariate analysis which is similar to factor analysis is Analysis of Variance (usually abbreviated to ANOVA). Although termed Analysis of Variance, perhaps confusingly what are actually being compared in the statistical test are the means of a number of variables, and ANOVA is used to compare more than two means at a time (Veal, 2011).

Case Study 8.2 is a discussion of the factor analysis and ANOVA that I conducted in relation to data that I collected at the Sidmouth International Festival. The factor analysis was an attempt to reveal if there were specific groups or subsets of attendees at the Festival, and if so what characteristics they had. The ANOVA was conducted in relation to the Likert-scale motivation statements presented in the questionnaire given to the sample of Festival visitors. Both of these statistical tests were also an attempt to compare the results I had obtained with the results from a previously published study, which investigated a similar type of music festival but in Australia (see Raybould et al., 1999).

Case Study 8.2: An example of factor analysis and ANOVA

A key aim of the Sidmouth International Festival research was to investigate what variables influenced visitor motivation for attendance at the Festival. The motivation factors presented to Festival visitors in the questionnaire are indicated in Figure 8.7. (here they are in the form of factor analysis results, but this was not the order on the questionnaire).

Prior to conducting the factor analysis, two important tests were conducted. One of these was the Kaiser-Meyer-Olkin (KMO) Measure and the other Bartlett's Test of Spehiricity (BTS). Both the KMO and BTS test indicated that the data were suitable for factor analysis. The factor analysis produced five groupings, though only three of these were fully acceptable in relation to the Cronbach Alpha test, which is normally acceptable at 0.7 and above. The factors are shown in Figure 8.7.

Figure 8.7: Factor Analysis of Sidmouth International Festival Data

Factor 1 (Cronbach Alpha 0.88) Factor Grouping: LEARNING

To be creative

To be involved and participate

To develop my skills

To challenge myself

To take part in musical activities

To interact with others

Factor 2 (Cronbach Alpha 0.80) Factor Grouping: FAMILY

To be with my partner/family

As I thought my family would enjoy it

To do things with my family

Factor 3 (Cronbach Alpha 0.69) Factor Grouping: PASSIVE ENTERTAINMENT

To be entertained

To experience an international festival

To watch musical events

Factor 4 (Cronbach Alpha 0.54) Factor Grouping: SOCIAL STIMULATION

To be with friends

As I enjoy the Festival crowd

Factor 5 (Cronbach Alpha 0.52) Factor Grouping: ESCAPE

To escape

As a break from my normal routine

It sounded like fun

There were three statements that did not correlate in the factor analysis: 'To expand my knowledge', 'The Festival is unique', and 'To meet new people'.

Note that each of the factors has been given a label – shown in capitals in Figure 8.7. These labels were very similar to, and in fact influenced by, the work of Raybould et al (1999) as well as other literature.

ANOVA was conducted in relation to the twenty motivation statements in the Likert scale. Statistically significant differences were found in relation to the following variables: gender, education level, marital/family status, nature of ticket purchased (season or day ticket), with whom respondents attended the Festival and whether respondents had attended the Festival before.

The results in relation to each variable, are discussed below in great detail. Females gave higher mean scores for eight of the motivation statements than males – these could be summarised as both 'push' and 'pull' motivational factors. In terms of education levels, those with higher educational qualifications produced results with higher mean results for the following statements: 'To be creative', 'To expand my knowledge', 'To challenge myself' and 'To be involved and participate'. Marital/ family status was important in terms of some results with family related motivation statements, perhaps not surprisingly achieving higher mean scores from 'family' Festival attendees. Those who responded in the categories 'Solo' and 'Solo/no children' produced results with higher means for the following statements: 'To be creative', 'To meet new people', 'To develop my skills', 'To challenge myself', 'To take part in musical activities', 'To interact with others'. Ticket type was important in influencing motivation. Season ticket holders gave responses achieving higher means for eight statements, the majority of which could be summarised as 'creative, active, learning, skill developing participation' at the Festival, while day ticket holders provided responses with higher means for 'family' related motivational factors. Those who attended with friends and family gave responses with higher means for family/friend related motivations. Those who had attended the Festival previously, in general, gave responses with higher means to the majority of the statements than those who were attending for the first time.

As should be clear from Case Study 8.2, using the statistical tests of factor analysis and ANOVA, although relatively complex and time consuming, can produce very interesting results indicating important details about the respondents in this particular research, as well as the differences between their views on a range of the topics in study. Hence, in summary, these forms of statistical analysis can reveal hitherto unknown detail and depth in relation to your results and if conducted appropriately should greatly enhance your study.

Writing this chapter in your study

An important reason that many students find that writing this chapter difficult, is that they appear to not understand clearly what the term 'analysis' actually means and therefore what processes they should be involved in, at this point in the research project.

At the beginning of this chapter, the question 'what exactly is analysis?', was presented and briefly discussed. As was indicated earlier, in the conducting of your analysis, you are attempting to explain your results, so that you can get a better understanding of them. Also as was also indicated, in attempting understand your results, the key process in which you are involved is comparison. Analysis is largely a process of comparison, and in your study will involve:

- comparison between the answers of one respondent to all of the questions on, for example, a questionnaire. In this way you will have achieved *a profile of this one respondent*, based on all their individual responses to your questions.
- comparison between answers given by different respondents to each of the specific questions on the questionnaire. This means that you will achieve *an understanding of your entire sample's answers to each of the questions*. When you have completed this for every question on the questionnaire, you will have an understanding of all respondents' answers to every question.
- comparison between your primary research and literature, including *a comparison between already published research on the topic you are researching, and your primary research.*

It is particularly the last point here (the comparison of your primary research with published literature) that many students seem unaware of or ignore when they conduct their analysis. However, if you consider that in most student research projects, one whole chapter is dedicated to literature, then you should also be aware that a key reason for this is so that you can highlight, initially, literature you will later refer back to in the chapter concerned with analysis. There is little point in actually having a literature review which does not present at least some literature that will be referred to again in the light of your primary results.

Analysis is the most important process that is required in your research project. The reason for this is that, through the processes of comparison outlined above, you will be attempting to answer the question 'What do my results actually mean?' It is in your analysis chapter that you will indicate to your supervisor and examiner whether you really understand what it is that your research has shown and your marks for your study will be greatly influenced by the quality of the analysis chapter.

The first stage of analysis is the preparation and presentation of your results, as discussed in Chapter 7. This should produce a summary of your results. However, be aware that a mere summary is not what is required in the analysis chapter. What is required here is your attempt to answer the vitally important questions: What do the results mean? To try to ensure you can explain what your results mean, the following questions should be helpful:

- Do you understand what your respondents have given you as answers? Can you communicate this clearly and convincingly in your chapter?

- Do you know what the responses mean? Can you communicate this clearly and convincingly?
- What patterns are there?
- What major and minor differences are there between responses from different respondents?
- If you have replicated a study, or your research is very similar to other research, how does your research compare with this?
- If you have used a number of different techniques (e.g. a questionnaire survey and document analysis) how do the results from each technique compare? What are the differences – what are the similarities? Do you know why the responses are different or similar? Can you explain this clearly and convincingly?
- How is your study linked to the wider literature in the field?

In terms of what you will actually need to do in your study is you should produce a discussion in which you provide the reason(s) why a respondent has given an answer to a specific question. Your reasons need to be based on evidence, not your guesswork. The evidence is quite likely to be found in responses not just to one question, but to a number of other questions you have asked in your survey, and/or in demographic information they have provided. For example, in my research into attitudes to a new tourism development in a rural area in New Zealand, I became aware that although three quarters of the respondents were in favour of the development a quarter were not. Within this quarter of the sample were a large number of female respondents, who were strongly opposed (according to responses to my Likert-scale statements) totalling just over 10% of the entire sample. I checked the demographic data and all the other responses of this group. This revealed that almost all of these female respondents lived close to the proposed development, which was a café/bar, were mothers in the age range 35-55 and several indicated in response to my open-ended questions that they had teenage children. Their actual responses to my closed-ended, thematic questions indicated they were concerned about increased noise, disruption to family life, drunken driving and accidents involving young drivers on rural roads. Using the evidence from both demographic data and thematic response, I felt I was in a strong position to explain the results of this group of female respondents.

The evidence to support your explanation of responses may also be found in literature. For example, when I conducted research into mountain biking in New Zealand, several of the female respondents indicated that they were unhappy with the provision of facilities for mountain bikers in general, and in particular for female mountain bikers. At the time of my research, New Zealand government documents indicated less than 10% of mountain bikers were female. So, in relation to the answers from female bikers, and what the government documents indicated about participation in mountain biking and current levels of provision of facilities, I could provide a convincing argument that there were, indeed, very few facilities for female mountain bikers and therefore they were justified in their responses in my survey about this situation. I could also back up my explanation with reference to academic literature, published at the time, which suggested a similar lack of facilities, for female mountain bikers, existed in many other countries, such as the UK, USA and Australia.

One reason that you are given several months to complete your study is to allow you the time to write and then reflect on what you have written in the early stages of 'writing up'. This is particularly important in relation to your 'Analysis' chapter. Your first attempt at analysis, for e.g. a questionnaire, may be to try to explain what individual respondents answers reveal about their views and understanding of topics and issues. Secondly it is likely to involve you in focusing on responses from all respondents to one specific question on your questionnaire. You will then probably work your way through all your question responses, taking each question in turn, until you have completed the entire questionnaire for all respondents. It may then be possible to write up your analysis on a question-by-question basis.

However, there may well be a point in the process of analysing each question in which you note that some of the responses are linked, or that certain key terms or phrases are important and keep recurring. You may then decide to discuss your findings, not question by question, but by topic, theme or issue. If you continue with this process of analysing using a topic approach, you may reach a point at which you realise that there is an overlap between some of your topics. So, the next stage will be to combine these topics to produce a smaller set of key topics that enable you to group almost all of your responses.

If you make use of statistics in your analysis, even if only descriptive statistics such as mean or mode, you need to provide a rationale for this. The ability to make relatively easy comparisons of results is quite likely to be, at least, partly the reason. There is even more reason to do this if you have used one or more statistical tests. You would need a very careful explanation of why this has been done, as well as an account of the actual processes you have used in the statistical test.

Student activities

1 What do you understand by the phrase: 'comparison is at the heart of good analysis' ?

2 Study Table 8.1, and on the basis of the data shown, write a paragraph that summarises the results of the Likert-scale questions.

3 What do you understand by the term 'statistic'?

4 What is a parametric test?

5 What is a non-parametric test?

6 Why is the term standard deviation important in quantitative data analysis?

7 What is the relationship between 'spread' and 'shape' in statistical analysis?

8 When would you use a Student's t-Test?

9 When would you use a Chi-squared test?

10 Case Study 8.2 presents data that has been subject to Factor Analysis and ANOVA. Why have these tests been conducted on this data and what do they appear to have revealed that was not obvious before the tests were conducted?

11 In relation to your own primary research, what is the main evidence you have to support your findings?

9 Qualitative Data Analysis

Introduction

There is a belief amongst some students that the data in qualitative research needs little analysis. This belief may stem from the idea that the data has been collected without reference to hypotheses and not within the confines of a researcher designed questionnaire survey, and this data is expressed solely in the words of the respondents. The belief leads to the conclusion that the responses should be allowed to 'speak for themselves' and should not be subjected to intensive scrutiny, interrogation and analysis. However, this is not the case. If anything, qualitative data needs more attention than quantitative data and may take longer to analyse, largely because the data has usually been produced in a much more open-ended context than quantitative data, so will be much more varied!

The discussion in this section assumes that data has been gathered in a largely open-ended way and that categories for analysis have not been set in advance, but actually emerge from the data, and that the material is being manually analysed, although there is discussion of computer programs that can assist with analysis towards the end of the chapter

The analysis process

Although you have begun the analysis process when summarising the responses to whatever qualitative research technique you have used, be it open-ended interview questions, focus group results or observational material, for presentation of your results, you have not at this point begun the process of explaining your data. Explaining the data is the key analysis activity.

Your first task will have been to try to bring some order to your results in terms of the content, so that you can prepare material for presentation, and this has been discussed in Chapter 7. In the process of preparing your results, you will have been familiarising yourself with the data, as this is the first step to summarising. This should then lead into the more complex stages of the process of analysis. However, part of this initial activity will be attempting to gain at least an interim understanding of your data. If you were involved in quantitative research, you may have set up a study which tests a theory, and so have an obvious way to analyse the material you have gathered, by comparing it to this theory. You may also have decided to use a particular statistical technique in advance of conducting your quantitative study. In this situation, you have a framework to assist in your analysis. It is very unlikely that you will be able to use these forms of highly structured and to some extent 'prepared in advance' analysis in relation to your qualitative data. But nevertheless, you will need to understand what your respondents have provided for you, via interviews, focus groups or observation.

The organising of your material may occur almost naturally, in that you will probably see patterns emerging as you read through transcripts of different respondents' answers. Certain words may be found in the responses, and you will begin to notice words that occur regularly and others that only come up occasionally, but still seem to be important. However, it may require you to note down what at, first glance, appear to be terms that occur often. Nevertheless, you need to be aware that terms that you find when reading through, for example, the first few interviews, may not occur in later ones that you read. So be aware that it is relatively easy to convince yourself that certain terms are being used by respondents, particularly if these are terms that you believed would be presented by respondents, before you started the analysis process, when, in fact, they are not being used as commonly as you believe. Noting down frequently occurring terms, as well as other terms that occur less frequently, will help you avoid forming an inaccurate impression of your data. This organising should enable you to summarise the data. This summarising process is very much an activity that you control but is based on what you have been given as responses, in that the summary terms are likely to be ones that have emerged from your data. A summary term may, in fact, be a concept, theme or issue referred to be a number of respondents. The key factor here is that these concepts, terms, topics and issues have 'emerged from your data'. Having said this, some qualitative researchers do set the categories in advance of conducting their primary research. They are usually experienced researchers who have conducted similar research before, albeit in a different context, and have a good idea of what to expect in terms of responses to their questions. In your case, as an inexperienced researcher, it is usually better to let your summary terms emerge from the data.

Remember, the process of summarising, is likely to be the initial stage in attempting to make sense of your qualitative data. This activity is, in effect, the first part of a key process in qualitative analysis. This will involve the coding and creating of categories in which to fit your responses. The coding or creating of categories is done partly to look for patterns in the data. The way these categories are created is via continually reading and re-reading the data. The ultimate aim is to create fewer and fewer categories. The process by which this is achieved is termed 'constant comparison' (see Glaser and Strauss, 1967). The term clearly suggests what the process actually is. The researcher reads and re-reads the data and the initial categories that have been created are, where possible, combined which reduces the total number of categories. There is usually a time gap between the initial reading and the re-reading, and in the gap, reflection takes place which should help with the reduction in the number of categories.

This process, of creating categories and then reducing them, is illustrated in Case Study 9.1 which is from my own research of primary school children's concept of a tourist.

Case Study 9.1: Children's understanding of the concept of a tourist

This case study is based on research I conducted as part of my PhD. The research evaluated a project focusing on the role of tourism as a topic in the UK primary school curriculum. Part of the research involved an investigation into what children had learned in the project. It had been my original intention to interview a sample of children aged 7-11 years old, outside their normal classroom environment. However, the class teachers indicated that this would be take too long and hence be disruptive to the children's learning. Therefore, instead, I used a questionnaire survey and this was conducted in class. The great majority of the questions were, unusually for a questionnaire, open-ended. Below I discuss just one such open-ended question and present a selection of responses to illustrate the process of constant comparison in refining categories.

The question was as follows: *What is a tourist?*

A selection of response is shown, to indicate how I went about creating categories and then refining them.

- A holiday maker
- A visitor to another place

- A visitor here
- Someone who comes here for a holiday
- A person who goes on holiday
- Someone who travels overseas
- Someone who stays here
- My gran' when she comes to stay
- Someone who leaves rubbish on the beach
- Someone who comes here from a long way away cos' they like our town

From these responses I considered possible coding or classification categories and initially came up with the following as my summary of children's understanding of the term 'tourist'. Note that, in my attempt to summarise, I use some terms that children themselves did not actually use:

- Non/resident - a visitor
- A traveller (domestic/international)
- Motivation - to take a holiday
- Accommodation - staying here
- Friend/relative (Visiting Friend/Relative category)
- Value judgements – polluter/admirer.

When I applied the initial coding categories to all responses, to this question, of the entire sample (the total was 152 children), it became clear that the term 'holiday maker' was used by 36% of the children, 'visitor to another place' by 28% of the children, and 'visitor here' also by 28% of the children.

Therefore, my final summary coding for responses to this question was as follows: 'The major components of the children's understanding of the term tourist is as follows:

- Holiday maker
- Visitor here
- Visitor elsewhere'

To emphasise, an important point in this process of creating categories is that you allow sufficient time. It is usually a very good idea to produce categories then leave these overnight (at least) before trying to reduce them in number. The concept of 'let's sleep on it' is therefore important here. Some researchers may put aside the data with its initial categorisation, for several weeks before looking again and re-reading and re-categorising. You probably do not have the luxury of this amount of time, but the more time you allow yourself the greater the likelihood of you feeling confident in your choice of categories.

Although you may feel confident when you have created summary categories, it may be worth getting another person to look at your data to confirm (or not) what you have decided.

The research presented in Case Study 9.1 involved the analysis of open-ended questions on a questionnaire and the reasons for the use of this technique are explained in the case study. However, recorded interviews can be transcribed and then analysed in very much the same way as the open-ended questions in the case study and there is detailed discussion of the processes involved in Case Study 9.4. In a similar way, transcribed recordings of focus groups can also be analysed following this process. It is also possible to analyse observations using a similar approach, as Case Study 9.2 indicates.

Case Study 9.2: Observation at Wellington Zoo

This case study is based on a research project I conducted at Wellington Zoo, New Zealand. I was investigating the various roles of the zoo and whether or not these were compatible. I was also interested in the nature of interpretation at the Zoo and how this related to its various roles. Observation was of the Zoo guides, or keepers, as they are known. The keepers gave a number of talks on different animal species. Talks concerning the following animals were observed: chimpanzees, 'big cats', Malaysian Sun bears, giraffes and exhibits in the Nocturnal House. As the observations were concerned with talks, listening as much as looking was a key part of the process. I did not create in advance a formal schedule of questions that I hoped would be answered, but I had decided, based on my research themes, to focus on the following:

- Keeper presentation style
- Keeper presentation content (particularly how much information was species specific and to what extent the content was more generally conservation focused)
- Audience reaction
- Keeper-audience interaction

My summary analytical notes were as follows:

Chimpanzees

The audience were an almost equal mixture of adults and children (mostly under aged 11) about 25 of them, about half of these being family groups. The keeper fed the animals during the talk and encouraged the audience to feed them, so interaction audience-animals. In fact, the process was incentivised, as the keeper asked questions and those who answered correctly were rewarded with a chance to feed chimps. Adults at least as keen to feed chimps as children were. The talk was mainly species specific, but information provided on differences between monkeys and other primates. More general conservation points raised in terms of habitat loss and discussion of endangered species.

Big cats

Observed twice. On each occasion tigers are the focus. First occasion – this is part of a 'Zoo School' activity. 24 children aged 9-11 the audience. Keeper feeds tiger by hand (outside cage) and discusses behaviour of tigers in the wild and in Zoo. This particular tiger was bred here in captivity, so a strong link to the Zoo/New Zealand. Children appear very impressed by being close to tiger and ask several questions, mainly species specific. Second occasion – Talk to general Zoo visitors (no feeding by hand). Animals very docile. Most of audience (mixed adults/children) look bored, fairly quickly. Research suggests for many zoo visitors, in general, the big cats are the highlight of a visit, probably not the case with this audience.

Malaysian Sun Bears

Observed twice. First occasion – large audience 60+ so amplification used. Audience comprises at least one third families with young (under 5) children. Talks last approximately 20 minutes, but some families with young children leave before half way through. Little keeper-audience interaction. Content largely species specific. Second occasion – smaller audience 25-30. Fewer parents with young children. Audience more attentive. More interaction keeper-audience. Focus of talk – bear cubs, as two have just been born and appear briefly from 'cave'. Talk lasts over 25 minutes, almost no-one leaves before completed.

Nocturnal House

Talks were detailed and informative, but almost exclusively species specific. However, the nature of the attraction means that it is dark and therefore difficult to hold the attention of the audience and some young children seemed more frightened than excited by the darkness. The species are from New Zealand and some only found in New Zealand, but even when awake they are not very active, so the talks are probably fairly short for this reason.

Giraffes

There is a viewing area, at the same height as the giraffes' heads, reached by steps. About half the audience were children aged under 11 and they were able to feed the giraffes assisted by the keeper and they appeared very interested in doing this. A small number of the audience left before the end of the talk, it seemed because they could not get close enough to the giraffes on the small raised platform. The content, in which there was much comparison of giraffes and humans, provoked positive responses and some questions from the audience.

I also produced summary notes based on all the talks, in which I used my original research themes. I discussed the overall content of the presentations, the quality of the keeper talks, the nature of audience, the reaction of the audience to the talks and the interaction between the keepers and the audience. I also referred to unexpected activities/events such as the keeper hand feeding the tiger, the significance to audience enjoyment of the sun bear cubs and the children's pleasure at feeding giraffes.

Content analysis

Earlier in this chapter there was discussion of the process of 'constant comparison' and examples from my PhD research in which I indicated how I analysed primary school children's responses to my question 'What is a tourist?' In this example I was analysing primary data. However, the process described and explained earlier can also be applied to secondary sources of material. In summary, this process is termed content analysis.

It is not that uncommon for students in their study to use content analysis of a secondary source as the main approach in their research and do it in such a way that they can argue that they are making an important contribution to our understanding. I will discuss below how I have used it in my research, but firstly will provide some more explanatory comments on the nature of content analysis.

Content analysis is a way of breaking down a large amount of data/material that is in a holistic format into smaller aspects that can be used to bring out more clearly, the 'hard facts' in the material (Clarke et al., 1998: 212). It was developed initially to investigate the content of newspapers and similar publications (Weber, 1990). The intention was to reveal how much of the publication was news, how much advertising, how much other material such as photographs, maps, cartoons and other illustrations. The analysis was further refined to break down news articles to investigate how much of the paper was devoted to particular news stories. This meant that the quantity of words in a story could be measured. Initially this was done via what was known as 'column inches'. This was literally the amount of measured space in the newspaper. The location of the story could also be investigated, in terms of whether it was on the front page or on inside pages.

A story could be followed over several days, or even weeks, to investigate how the amount of space devoted to it and the story's location in the paper changed over the time period.

This measuring of the number of words and space devoted to a story indicates an important dimension of content analysis. It is a process that studies largely qualitative material but creates numeric categories. In other words, it can turn the qualitative into the quantitative! The fact that the initial focus of much content analysis was on newspapers is also an indication of how it can be used today. The source of news stories has changed greatly over the past one hundred years – now radio, TV and increasingly the internet are the major sources of news stories. However, content analysis can also be applied to audio-visual material. The most common way of measuring will be the length of time of a story. But it could also be, for example, what proportion of a thirty-minute TV broadcast or radio broadcast has been devoted to the story, or it could be where in the running order a particular story is featured. It could also be whether the original story is backed up with discussion/analysis by media commentators/experts.

My research into the media concern with the SARS outbreak in early 2003 and its link with tourism made use of content analysis and involved a conventional approach to the analysis (see Mason et al., 2005). Severe Acute Respiratory Syndrome (SARS) first occurred in mid-February 2003. It would be more accurate to indicate that this was when it was recognised by the global media. It had actually been in existence since the previous November. The significant tourism dimension to the potentially lethal disease was that having originated in SW China, and then been taken accidentally to Hong Kong, it was spread, very rapidly, by tourists, to a number of locations around the world, along major air routes. The media at the time of its occurrence (particularly newspapers, but also TV, radio and the internet) devoted a large amount time/space to the SARS story. My interest was in the following: how much space was devoted to SARS in selected newspapers and on what page were their stories; how much time was given to SARS on TV and where in a news broadcast was the story; what was the prominence of SARS on the internet; for how long the SARS story remained a major news item and; whether or not the nature of the SARS story changed over time.

In terms of the analysis process, my research team and I selected a number of newspapers in the UK, NW Europe, Asia, the US and Australia and, for each, studied the content of the SARS story, the location in the newspaper (the page numbers) and how much, as a percentage of the paper, was devoted to the SARS story. We did this for a six-week period immediately after the reported outbreak in February 2003. We also viewed a number of TV channels, broadcast from within the UK, and also selected TV channels based in the US, Asia and Australia on the internet. For this media, over the same six-week period, we noted the length of time given to the SARS story and at what point in selected news bulletins it featured. In addition to collecting data from these different sources, 'the issue-attention cycle' theory (see Hall, 2002) was applied to the SARS outbreak and its media coverage. This theory suggests that media stories follow a cycle in which they come to prominence and large amounts of media time are devoted to them, they reach a peak of interest, before fading away as other stories take over and then dominate the news. Amongst a number of findings, the study indicated that the media focused on the 'shock/horror' element of such a story, and tended to over-exaggerate certain aspects, but also that the issue-attention cycle theory was found to largely fit the events surrounding the SARS outbreak.

Content analysis can be particularly useful in other ways in tourism and related fields. Much tourism material produced (particularly that used in marketing) uses several visual stimuli, as well as words. This type of material can be studied using content analysis. So, for example, it is possible to conduct analysis into tourist brochures, as is indicated in Case Study 9.3.

Case Study 9.3: Content analysis of a cruise ship brochure

The cruise ship brochure (the one used was Fred Olsen Lines 2007, but could be almost any major cruise ship brochure) was not analysed in terms of its word content, but solely the visual images. The brochure had the following images:

Photographs of:

- Ports/ locations where the cruise ship visits during particular cruises
- The exterior of the cruise liner while at sea
- Examples of the interior of different quality/class of cabins
- Facilities on board, the bar, performance stage, ballroom, swimming pools, fitness areas, shops, restaurants
- Examples of typical customers
- Examples of crew members.
- Maps of the route taken for various cruises.
- Climate graphs for different locations.

The analysis of the brochures was solely in relation to the visual images – all words were ignored. However specific questions and topics were used to analyse the images:

1 Who are the tourists, in terms of e.g. age, ethnicity, whether single/couples/families?
2 Who are the crew in terms of e.g. age, ethnicity?
3 What do the crew do?
4 What do the tourists do?
5 What interaction is there between tourists and crew?
6 What information do the climate graphs show?
7 How are the climate graphs linked to the other visual material?

Although Case Study 9.3 indicates only a focus on the visual material, there are of course opportunities to analyse the written content of material such as tourist brochures. It would be possible to analyse both the written content and the visuals of one specific or several different brochures. It could then be possible to compare and contrast what is indicated by these two ways of presenting with brochures, and this comparison could form an important element of analysis within your research project.

Whether visual or written material, Case Study 9.3 discusses content analysis of what has been printed, and in this case, as indicate above, I used tourist brochures. There are other types of tourist material that could be analysed such as guidebooks. Much tourism promotional material is now found on the internet, as well as in printed form. This material could be very suitable for content analysis. The type of questions/approaches to analysis would need to be decided on but could be similar to the types of questions posed in the case study.

However, it is very likely that you have collected primary data in your qualitative study and interviews are a frequently used technique to collect this qualitative data. Case Study 9.4 indicates in detail the processes involved in analysing qualitative primary data and is based on the work of a PhD student. The study involved in-depth interviews with tourists in relation to their understanding of the meaning of the concept of tourism destination quality.

Case Study 9. 4: Data analysis for the qualitative phase of the Tourism Destination Quality (TDQ) study

The goal in qualitative data analysis is to achieve results that closely reflect the respondents' opinion on a phenomenon, which is the researcher's focus of enquiry. To achieve its results, qualitative data analysis uses inductive logic wherein patterns, themes, categories or relevant variables come from the data itself, rather than being imposed on them prior to data collection and analysis. Therefore, the approach to analysing qualitative data can be viewed as having a goal to 'let the data speak for itself', which enhances the ability to obtain results that are grounded in the data. In the first exploratory qualitative phase of the TDQ study, one of the widely used inductive methods for analysing qualitative data, the constant comparison method (Glaser & Strauss, 1967), was employed to analyse the in-depth interview data.

There were five major steps involved in analysing data using the constant comparison method, based on Maykut & Morehouse, (1994) which were as follows:

1. Preparing the interview data for analysis
2. Unitising the interview data
3. Developing and assigning 'units of meaning' to categories
4. Transforming the 'units of meaning' to attributes
5. Refining categories and attributes.

Step One: Preparing the interview data for analysis.

The in-depth interviews were transcribed and the data transcripts appropriately labelled. Labelling involved tagging each data transcript with a unique serial number, written on the top right-hand corner of each data transcript, indicating when and where the in-depth interview was conducted, and on what page of the data transcript the interview could be found. For example, the labelling tag D1/01/p1 denoted the first in-depth interview (D1) conducted on the first day (01) of data collection, which can be found on page one (P1) of the transcription text. Labelling data sources enhances the creation of an audit trail that other researchers could follow if they needed to retrace the original researcher's footsteps. In turn, creating an audit trail in qualitative data analysis facilitates easy external scrutiny, which enhances the credibility of both the data analysis process itself as well as its results. Subsequent stages of the data analysis required certain sets of the data from transcripts to be cut out. As a result, a mechanism to preserve the original transcripts for future reference was employed. After labelling, the transcripts were photocopied page by page and the original transcripts were placed in marked envelopes and stored for safekeeping. Photocopies of the transcripts were then used in subsequent stages of data analysis.

Step Two: Unitising the interview data.

The next step in the data analysis process was to unitise the data. This is the process of searching for meaning in the data through identifying '...chunks or units of meaning in the data'. 'Units of meaning' are words, phrases or paragraphs representing a single meaningful idea. In the initial stages of the data analysis process, small 'units of meaning' in the data were identified. These small units of meaning would then form the basis for defining larger categories of meaning in subsequent stages. The process of unitisation is consistent with the overall process of analysing qualitative data, which has often been described as one of: '...culling for meaning from words and actions of the participants in the study, framed by the researcher's focus of inquiry' (Maykut & Morehouse, 1994: 127). Lincoln & Guba (1985) suggest a number of iterative steps for the unitising interview data, which were used here. These are as follows:

1 Familiarisation with each interview;
2 Identification of 'units of meaning' from the data;
3 Extraction of 'units of meaning' from the transcripts.

Familiarisation with each interview. Before the process of unitisation could begin the researcher had to be well acquainted with the data to be analysed. This was achieved through playing back and listening to the interview recordings and reading the interview transcript three to four times.

Identification of 'units of meaning' from the data. Once familiarised with the data, the next step in the unitisation process involved searching for 'units of meaning' in the data set and this was again meticulously read but now with a different aim which was to identify 'units of meaning'. According to Lincoln and Guba (1985), in order to be useful for further analysis each 'unit of meaning' should stand by itself, that is, it must be understandable without additional information. Once a 'unit of meaning' was identified a line was drawn across the page to separate it from the next 'unit of meaning'. A labelling code was placed in the right-hand margin, indicating where the 'unit of meaning' was to be found. This code ensured that the 'unit of meaning' could always be traced to its original source to provide the contextual detail that may have become necessary as the data analysis proceeded (Hewitt-Taylor, 2001). In addition, the labelling code also served to create an audit trail. The quality of data analysis, using the constant comparison technique is dependent on the extent to which a repeated, systematic search of the data is conducted (Hammersley, 1981). In this exploratory study, there were repeated searches for 'units of meaning' in the transcripts until no new insights were discovered.

Extraction of 'units of meaning' from the transcripts. Working with copies of unitised transcripts, each 'unit of meaning' was cut out and pasted on an index card; one 'unit of meaning' per index card. The process resulted in a total of 175 index cards, each containing a 'unit of meaning'. At this stage the 'units of meaning' were ready for further analysis.

Step Three: Developing and assigning 'units of meaning' to categories.

The next stage involved developing categories and assigning 'units of meaning' to categories. To begin this process, the researcher assembled the 175 unitised index cards, several blank index cards and a large sheet of paper that were pasted to the wall. Then, a number of steps were followed to complete the process of developing categories and the assignment of 'units of meaning', which included:

1. Development of the first category and
2. Development of subsequent provisional categories.

Development of the first category. The researcher picked one of the cards containing a 'unit of meaning', read it and then developed the first higher-level category that summarised the meaning of the 'unit of meaning' conveyed. Therefore, consistent with the inductive approach to data analysis adopted, categories were developed, *in vivo*, i.e. from the data itself rather than being pre-determined. The first provisional category was written on a blank index card and pasted on the left-hand side on the paper. The index card, which had led to the development of the first provisional category, was pasted underneath it. The remaining unitised cards were reviewed to see if any of them fitted the newly created first category. If a second data card that seemed to fit the first category was found, a decision on whether or not to include it in the first category was made, based on whether the meaning on this second card 'looks like' or 'feels like' the one in the first category. If the second card was found to fit the first category, it was pasted underneath the first card. The process was repeated until a card, which did not fit the first category, was picked. This led to the next step in the data analysis process.

Development of subsequent provisional categories. When a card was found that did not fit the first category, a new category was developed using the same process that resulted in the development of the first category, outlined above. The new category was written on an index card and pasted to the right of the first category. The unitised card that led to the development of this new category was pasted underneath it. The remaining cards were examined to see if any others fitted this new category, based on the 'looks like' or 'feels like' criteria. If another card, that fitted the new category, was found it was pasted underneath. The process was repeated until all unitised cards were allocated to a category. However, the 'units of meaning' for each provisional category were still, at this point, in 'raw form', i.e. expressed in the words of the interviewee and needed converting into attributes of tourism destination quality (TDQ).

Step Four: Transforming the 'units of meaning' into attributes.

Transforming the 'units of meaning' into attributes of TDQ involved a systematic process whereby the researcher worked with one category at a time. To create the first attribute, the researcher picked an index card consisting of a 'unit of meaning' belonging to a specific category and then read it. Using wording as close as possible to that in the 'unit of meaning', the researcher formulated the first attribute of TDQ for that particular category dimension. The attribute was typed into a Word document under the specific category to which it belonged. To develop a second attribute of TDQ, the researcher picked a second 'unit of meaning' from the same category as the first 'unit of meaning', formulated the second attribute and typed it under the first category where it belonged. This process was repeated until all the index cards falling within a given category had been exhausted, before moving on to the next category. The same process was repeated in transforming all 'units of meaning' falling from other categories into attributes of TDQ. This process resulted in the initial list of attributes being allocated to respective dimensions, and ready for further analysis.

Step Five: Refinement of categories and attributes.

The next stage in the data analysis process involved refining the categories and attributes of TDQ. At this stage, the researcher decided to invite two tourism academics, each with a postgraduate degree in tourism and not connected to the study, to: a) critique the categories and attributes developed by the researcher and b) suggest, if necessary, the categories or attributes which could be further merged so as to try and eliminate duplicates. Involving researchers to critique, comment or make suggestions during qualitative data analysis is a widely employed strategy for bringing external scrutiny to the data analysis process, thus enhancing the validity of the outcomes of the study. The academic tourism experts were each given a copy of a list of attributes of TDQ, grouped under the categories developed by the researcher. Working separately from each other, the two tourism experts made suggestions regarding categories and attributes they thought could be combined to avoid duplication. On the basis of the two experts' comments, the researcher merged the categories and attributes that seemed too similar to remain discrete and then produced a revised list of the categories and attributes of TDQ and gave these to the experts to review again, and they made further comments on the revised list of categories and attributes. The researcher revised the list of attributes again. At each stage, the researcher discussed with the experts the changes they were recommending. The process was repeated until no further revisions were recommended. The final result was a list of attributes and dimensions of tourism destination quality.

(Source: Seakhoa-King, Augustyn and Mason, 2021)

Case Study 9.4 provides details on how to conduct analysis of transcribed interview material and how this leads to the generation of categories and of a number of attributes. Case Study 9.5 indicates how attributes, generated via the interviews in Case Study 9.4., were grouped into dimensions. This process of grouping the 75 attributes in the TDQ study resulted in the creation of 12 dimensions: affordable, authentic, safe, well-kept, hospitable, informative, novel, varied, relaxing, uncrowded, all weather and child-friendly. Case Study 9.5 provides just two of the dimensions and the related attributes to indicate the links between attributes and dimensions. These dimensions are 'Hospitable' and 'Safe' and in each context the attributes are listed in order of the number of times they were indicated in the interviews by respondents, with the most frequently stated at the top of the list and progressing from here down the list of attributes in descending order.

Case Study 9.5: Dimensions with their attributes in the TDQ study: two examples 'Hospitable' and 'Safe'

Dimension: **Hospitable**

Attributes:

'Welcoming local people'

'Local people who are willing to advise tourists on the local area'

'Local people who are willing to help tourists'

'Local people who encourage tourist to become involved in local activities'

'Local people who make tourists feel like locals'

Dimension: **Safe**

Attributes

'No danger of facing physical harm'

'A low crime-rate'

'No political unrest'

'No risk of being mugged'

'No risk of being overcharged'

'No danger of verbal abuse'

'Security for personal belongings'

'No beggars on the street'

'A visible police presence to assure tourists'

(Source: Seakhoa-King, Augustyn and Mason, 2021)

Computer analysis

It may at first sight seem unlikely that you can analyse qualitative material using a computer as it is not likely to be in numeric form. However, there are now a number of extremely good software packages that do just this. Many will sort and code qualitative data, and analyse non-textual material such as visual or audio material, and can store data. One major qualitative approach is computerised content analysis. The key advantage of the use of software to conduct content analysis is that the computer is not subject to the type of 'quirky' behaviour that some people reveal when they manually analyse content. The discussion NUDIST explains what I mean by this.

NUDIST

NUDIST was one of the earliest packages for content analysis. An interesting name, 'NUDIST' stands for Non-Numeric Data, Searching, Indexing, and Theorising. The key features of this computer package, which has been available for approximately thirty years, are that it documents, indexes and then analyses unstructured qualitative data. If a researcher has a 'hunch' that certain words or phrases are occurring frequently when dealing with a large amount of qualitative data, NUDIST will be able to confirm (or reject this) as it will search for and count occurrences of the words/phrases. It will also do what is not always done when one person is analysing material on their own manually – it can spot and do a frequency count of words/phrases that have been missed or overlooked by the 'quirky' researcher! It is important to note here that the researcher needs to input words/phrases for the computer to search for – the computer does not think of its own accord.

ATLAS ti

'ATLAS ti' is another, although more recent, computer package for analysing qualitative material. It works in a similar way to NUDIST and enables interviews, focus groups and field notes to be analysed. It comes with tutorials and there is also online support. ATLAS ti is also able to analyse non-written/type-based material, so it can be used with audio and video material. It can also analyse geographical material such as maps, as well as can be used with photographs and internet images.

NVivo

Another popular computer package for qualitative data analysis is NVivo. Like NUDIST and ATLAS ti, NVivo is able to analyse written or type-based material, but its more recent versions can also analyse social media, such as Facebook, YouTube videos and webpages.

Alceste

Alceste is a computer package that is particularly useful for analysing a whole body of work, which could be in the form of a document, as well as primary data, as long as it is in text format. It can also be used to analyse a group of, for example, interviews that contain a number of similar/related topics/themes. The package analyses the vocabulary used and classifies segments of text/phrases, where there are similarities. It is able to convert qualitative data and make it possible to conduct statistical analysis.

NUDIST, Atlas ti and NVivo are three computer packages that are used widely in academia. There are several more that you may wish to find out about for yourself. However, it is very likely that a part of the institution you are linked to will have copies, or at least a licence giving permission to use, NUDIST, Atlas ti and NVivo. A great advantage of these computer packages is that it is possible to use them in conjunction with other non-qualitative data focused computer software, including Microsoft Word, Microsoft Excel, SPSS, EndNote and Survey Monkey.

Computer packages such as those discussed above can therefore be used to help conduct the type of analysis discussed in Case Study 9.4. which is actually an example of manual analysis. Using a computer package should greatly speed up the process of manual qualitative data analysis. One way it can do this is to identify all sections of text, such as from an interview or a focus group, which has been coded in a particular way (see Veal, 2011), such as that described in Case Study 9.4. Strictly speaking this identification is not actually analysis, but preparation for analysis, as the computer cannot provide an explanation of why, in the case of, for example, interviews, respondents gave the answers that they did, but only identify sections that have the same coding. However, this should help prevent the type of 'quirky' coding referred to above.

Writing this chapter in your study

To help you get started on the process of making sense of your qualitative data, these questions should give you focus:

- How did I collect this data and how has this affected what I have collected?
- What data have I actually got?
- What are the more important and less important topics in the responses?

- What patterns in the data can I see?
- Who else would be interested in the responses and how could this material be used beyond my dissertation?

These questions do not have to be answered immediately at the outset of analysis but should continue to guide you during the process of analysis.

Prior to actually writing this chapter you need to have a good grasp of your data and what it means. As indicated earlier in this chapter, the creation of categories based on the data you have collected is an important early activity contributing to the analysis process. This will involve reading and re-reading your source of data, be it interviews or focus groups or possibly field notes based on your observations. A good way to carry out this process is to make notes on your transcript each time that you read it. This may result in the repetition of category names, but it could also lead to new categories. These new categories may initially appear to be adding to, and not reducing, the number of categories. However, it is often the case that the new term you have used to categorise is actually a broader concept, which is likely to encompass several of the existing categories you have created. So, in this way it is actually reducing the number of categories!

However, the most important part of the process of analysis is when you are trying to answer the question 'why' – particularly the following 'why' questions:

- Why have I achieved these results?

But this also should lead to the more specific 'why' question:

- Why have these particular respondents given me these specific answers?

These are the type of 'why' question that you are very likely to be faced with and they need to answered, if you are to conduct high quality analysis and, hopefully, achieve a good mark in your project. It may not surprise you to be informed that answering these types of question is probably the most difficult part of your analysis and in fact the whole research project. How to conduct this part of the analysis is very important.

Using the material in Case Study 9.1, involving children's understanding of the term tourist, previously referred to, I indicate below how you can carry out this type of analysis and answer the question: 'Why did my respondents give me these particular responses?'

As indicated in Case Study 9.1, the question presented to the primary age children was: 'What is a tourist? The responses were initially categorised and then reduced to just three groups. The categories were: 'holidaymaker', 'visitor here' and 'visitor elsewhere'. So, at this point, the key analysis process is to seek the answer to the question: 'Why were these the major categories of responses?'

In attempting to answer the question 'why', a very important aspect is the context of the questioning. First, the specific research instrument and the way it was used were important here. So, the questioning, although open-ended, was restricted by the length of time available and the fact that the respondents were relatively young (7-11 years old) primary school children. Also, the children were required to write their responses, rather than say them – the original intention had been to use interviews, rather than an open-ended questionnaire. Using these combined factors means that it is possible to suggest why the children's responses were not lengthy.

In terms of the content of the responses, again the context of the research, both in terms of the nature of the research technique and how it was actually used, was important. The research was conducted in schools in, what was at the time, the most important geographical region in the UK for domestic tourism. Therefore, it is possible to claim that even young children would have been aware of tourists in this region. Hence, it is perhaps not surprising that children were aware that the term 'tourist' is another word for a holiday maker. In relation to the range of answers to this question, it would be appropriate to compare the responses of different children involved in this research, some of whom will have given the responses indicated in the initial categorisations presented in the case study, but also others. Verbatim statements could be used to exemplify children's comments as well as summaries in the researcher's words. The children' responses could also be compared with literature, particularly literature concerned with meanings and definitions of the term 'tourist'.

An important point for the analysis, is that the context of the research is also likely to be the reason a relatively large number of children referred to 'visitors here'. Such responses would probably not have occurred from children in an area of the UK where relatively few visitors were evident. However, without further research, this cannot be known for certain. Nevertheless, the original research could be followed up by more specific questioning of the children who were involved initially and additionally new research, with similar types of children, could be conducted in other parts of the UK where there were relatively few visitors. This type of research could be suggested as a recommendation, indicated by, and following, on from the analysis.

So, in summary, in my analysis, what reasons did I provide for why I achieved these results? These are indicated below:

- The research instrument used – an open-ended questionnaire which respondents were required to answer in a given space.
- The nature of the respondents – school children aged 7-11
- The geographical location/context – Sout-West England, where domestic tourism is important
- The setting for the research – at school with teachers influencing my choice of research approach and technique.

Finally, a key message to assist when writing this chapter in your study – it should be clear from this discussion, that evidence from the research has been used to attempt to answer the question 'why?' and this has not involved 'guesswork'.

Student activities

1 What are the reasons for looking for patterns in your qualitative data?

2 Explain the term 'constant comparison' and indicate how you would use it in relation to your qualitative data.

3 In relation to Case Study 9.2., assume you are conducting research, not at a zoo, but at **either** a one-day music festival **or** a sporting event taking place on just one day. Indicate what you would observe at your specific activity and create the criteria that you would use to assess what actually occurs at your event. How would you analyse your data?

4 Study Case Study 9.3 and in particular the criteria used to analyse brochures. Now select a country or region's marketing website of your choice. Indicate what criteria you would use to conduct a content analysis of the site.

5 Case Study 9.4 provides a detailed account of five major steps involved in analysing data using the 'constant comparison' method

   a) Explain how the process of constant comparison has been used.

   b) Write a short summary (maximum 200 words) of the processes involved in the five steps

6 Case Study 9.5 indicates how attributes (derived from the processes indicated in Case Study 9.4) were used to create dimensions. Explain how this was done, in your own words.

10 Analysing Mixed Methods Research

Introduction

As discussed in Chapter 6, mixed methods research will involve some form of combination of quantitative and qualitative techniques. It would therefore seem obvious that the nature of analysis within mixed methods will involve the combining of the processes discussed in relation to quantitative and qualitative research. To a large extent this is the case, but can be somewhat more complex than this.

Mixed methods research analysis

An important reason that mixed methods is not necessarily as straightforward as just combining quantitative and qualitative analysis is that, as noted in Chapter 6, there are several different types of mixed methods research, with Creswell (2009) suggesting there are six types, and two major groupings in this sixfold typology, and these depend on whether the methods are used *sequentially* or *concurrently* (Creswell 2009).

As outlined in Chapter 6, in the mixed method grouping 'sequential' research, one approach commences with qualitative research and is followed by a quantitative phase, and this is *sequential exploratory research* (Creswell, 2009). Research beginning with a quantitative approach, followed by a qualitative phase is referred to as *sequential explanatory research* (Creswell 2009). The third type of sequential research uses a specific theoretical perspective at the outset, which shapes the direction of the research. Here, the sequence of the two-phase research can begin with either a quantitative or a qualitative approach and Creswell terms this *sequential transformative research*.

As Chapter 6 indicates, in the second major grouping, concurrent research, there is *concurrent triangulation*, where qualitative and quantitative data are collected concurrently, and here equal weighting is given to each of the qualitative and quantitative approaches (Creswell, 2009). However, in the *concurrent embedded approach*, there is one predominant approach, with the secondary supporting approach, (which can be quantitative or qualitative), embedded within the main approach. The last approach is the *transformative concurrent approach.* where the collection of both quantitative and qualitative data is guided by the use of a specific theory (Creswell 2009).

In relation to sequential mixed methods the process of analysis will be largely guided by whether quantitative or qualitative methodology was used first, and which came second. This will be so because it is very likely that the results of the first methodology used will have influenced precisely what was done in the second stage of research, when the other was used. Hence, if the first methodology was qualitative and made use of interviews, it is likely that the results from these interviews have helped in the second quantitative phase. If a questionnaire was used in the second phase, then interview results will probably have helped create questions that have been posed on the questionnaire that was used in the second phase of research.

Case Study 10.1 provides a discussion of the processes involved in an example where qualitative research preceded a quantitative phase. This case study is closely linked to Case Study 6.2. in Chapter 6, where the processes of design and conduct in the qualitative phase of mixed methods research are discussed and also to Case Study 9.4. where the process of analysis of the qualitative phase of the research is discussed. In Case Study 6.2 the details of the decision-making process of the exploratory qualitative research phase are discussed and there is brief reference to how this was linked to the second quantitative phase. However, the results from the first phase are not presented or discussed in Case Study 6.2. Case Study 10.1 shows how the results of the first qualitative phase were analysed and then contributed to the design of the research in the second quantitative phase.

Case Study 10.1: Tourism Destination Quality: Analysis of sequential exploratory mixed methods research

As Case Study 6.1 indicates, the research involved was an investigation into what tourists understand by the concept of tourism destination quality (TDQ). As it was considered that little was known about tourists' understanding, in the first stage of the research a qualitative approach was used. This involved the use of in-depth interviews, which were initially pre-tested and then piloted before a total of 41 interviews were conducted until a saturation point had been reached.

The analysis of the results of the interviews using 'constant comparison' (Glaser and Straus, 1967) led to a number of attributes and dimensions associated with tourism destination quality and the step-by-step processes by which the data from the interviews was analysed is discussed in detail in Case Study 9.4. In addition, several initial categories that were generated in the analysis were reviewed and a number of them subsequently merged. However, in summary, the analysis revealed 75 attributes of destination quality which were then grouped into 12 dimensions.

Having obtained results from the first qualitative phase of research it was now possible to consider what to do in the next phase. A decision was then made to follow up the qualitative approach with a quantitative one. The use of quantitative research in the second phase was intended to provide an opportunity for findings from the qualitative phase to be assessed and evaluated, using a larger, more representative sample of the population of tourists. Also, it was believed that a quantitative approach would enable an assessment of whether there were any significant differences in responses among tourists, based on their socio-economic and travel characteristics. Therefore, a quantitative survey research approach was adopted which involved collecting data about respondents' characteristics and opinions.

As the intention was to create generalisable findings from a larger, representative group of respondents than those involved in the qualitative phase, the quantitative research technique deemed appropriate was a questionnaire survey. Therefore, the attributes obtained in the qualitative stage were turned into statements which were inserted into a printed questionnaire. These were set within a 7-point Likert-scale, where '1' was 'Strongly Disagree' and '7 'Strongly Agree'. These statements became the key stimuli as questions on the survey used in the second phase, along with closed-ended questions focusing on respondents' demographic and travel characteristics.

The questionnaire was pre-tested using students from the University Business School, then piloted at a UK shopping mall involving 40 tourists, with minor modifications to the closed-ended questions being made. The revised questionnaire was then distributed at two airports (London Luton and London Stansted) which were regarded as 'neutral' locations – i.e. not specific tourism destinations. A total of 806 questionnaires were distributed and 752 completed, usable questionnniares were obtained.

The respondents' score for each of the 75 statements had to be in the range 1-7, because of the use of the 7-point Likert scale, and descriptive data analytical techniques using SPSS for Windows were employed to indicate which attributes and dimensions tourists most strongly associated with TDQ. The most commonly used descriptive statistics, mean and standard deviation scores, were calculated for each attribute and dimension, and then ranked in a descending order, based on mean scores values. The mean score values were interpreted as follows: the lower the mean score, the less tourists were considered to associate that particular attribute, or dimension, with destination quality, whilst the higher the mean score, the greater the extent to which tourists were considered to associate it with destination quality.

Statistical data analytical techniques for comparing groups were employed to investigate a number of hypotheses that had been established on the relationships between dependent and independent variables. The dependent variable was what tourists associated with TDQ (represented by the twelve dimensions), while the independent variables were the tourist demographic factors such as age, gender and education level. The main goal for investigating these hypotheses was to establish whether there were any significant differences in tourists' responses based on their socio-economic and travel characteristics. Two types of statistical data analytical techniques for comparing groups were used. They were: (a) tests to establish whether groups were significantly different, and (b) tests to ascertain the strength of association between the dependent and independent variable. As the main questions had used a Likert Scale it was possible to conduct both parametric and non-parametric statistical analysis.

The main reason for using both types of test is that there is debate about whether self-administered questionnaires using a Likert scale, meet the requirements of parametric tests, as some researchers argue Likert scales use an 'ordinal' scale, whilst others that they use an 'interval' scale. Given the significant amount of controversy, both parametric and non-parametric tests were employed. However, parametric tests were the primary statistical techniques on which the interpretation of the results of this phase were based. Depending on the hypothesis being investigated, the following tests were used to establish whether groups were significantly different: t–test for independent samples, the Mann–Whitney U test, the one-way between-groups 'Analysis of Variance' (ANOVA) and the Kruskal Wallis test.

As the t–test and the ANOVA test can only reveal whether group differences are statistically significant, and do not provide any indication as to the magnitude of the difference, it was necessary to conduct additional tests to assess the meaningfulness of such differences. Such information is particularly important given that small differences can be statistically significant, especially where relatively large samples are involved. One way of assessing the meaningfulness of statistically significant findings is to calculate the 'effect size'. This is a set of statistics which indicate the relative magnitude of the differences between mean score values. There are a number of different 'effect size' statistics that can be computed. In the TDQ study, a frequently used 'effect size' statistic, *eta* squared, was employed to ascertain the strength of association between the dependent and independent variables.

The analysis in the quantitative phase of research led to the conclusion that the results of this phase were representative of the whole population of tourists and also revealed which specific attributes and dimensions tourists most strongly associate with destination quality. In descending order of importance, according to their mean, the dimensions were as follows: Authentic, Safe, Well-kept, Affordable, Informative, All-weather, Uncrowded, Hospitable, Relaxing, Varied, Novel and Child-Friendly. Of the 12 dimensions, only 'Child-Friendly' had a mean value less than 4 (the midpoint of the Likert scale) and the top eight dimensions each had a mean of over 5. The statistical analysis of the socio-demographic factors and travel characteristics of respondents investigated in the quantitative phase indicated that tourists do not substantially differ in relation to their assessment of destination quality. At this point it was possible to begin the comparison of both phase one and phase two results and generate overall findings. These confirmed that the findings of the first qualitative phase of research were very largely supported by the quantitative second phase.

Case Study 10.1 is an example of sequential, exploratory, mixed methods research and the overall results indicate the added value that using mixed methods can bring to a research project, particularly in terms of the confirmation in the second phase of what had been found in the first phase. As Teddlie and Tashakkori (2009) state, there are two other types of sequential, mixed methods research and Case Study 10.2 provides details on sequential, explanatory research, where a quantitative phase preceded a qualitative phase.

Case Study 10.2: Mountain biking in New Zealand: An example of sequential explanatory mixed methods research

This case study is concerned with the sport of mountain biking and centred on a town in New Zealand. The design of the qualitative part of this research (the use of focus groups) has been outlined in Case Study 5.4. Nevertheless, it was not discussed in detail there that it was part of mixed methods research. However, working in partnership with the local town council, the two main researchers involved were attempting to find out mountain bikers' attitudes towards and satisfaction levels with facilities in the town, and what improvements, if necessary, could be made in the future.

The first stage of the mixed methods research was a questionnaire survey sent to all members (totalling 136 at the time of the research) of the mountain biking club in the town. This asked members their experience of mountain biking, their views on facilities that existed then and what changes could be made to improve mountain biking in the area. This was part of the local council's attempt to gain data to contribute to its leisure and recreation plan. This initial phase of research involved a largely closed-ended questionnaire, which included a number of 'Yes/No'. 'Tick-box' and 'Likert-scale' questions. Analysis of results, which was conducted manually, revealed a growing interest in mountain biking in the town, but also a concern about the lack of funding and facilities to support it. Respondents were asked on the questionnaire to indicate whether they were prepared to attend a focus group following completion of the questionnaire, to further explore issues of mountain biking in the town.

As many as 80 members of the club responded to the questionnaire and a total of 23 indicated that they were willing to attend a focus group. This number of potential focus group attenders led to the decision to have two (one hour) focus groups – arranged on the same day with one following immediately after the other. The intention was to have 12 attenders in one group and 11 in the other, although on the day all 12 of those who been invited came to the first focus group, but only 4 arrived for the second one.

The questions used in the focus groups were based on responses given in the questionnaire survey, which had been analysed manually, as the sample was relatively small. One of the two researchers acted as the group coordinator and led the questioning, whilst the other researcher took notes on the ensuing discussions. Initial focus group questions asked for confirmation of views stated in the questionnaire, although these were not directed at specific individuals' responses as indicated in their completed questionnaire, but were posed to the whole group. However, most of the questions used in the focus groups asked for explanation of some of the answers given in the questionnaire. They began with phrases such as: 'Could you tell us why.....' Several other questions, which were posed towards the end of the focus groups were seeking new information from respondents.

Each focus group was recorded using an electronic device but also one of the team of researchers kept a written record of the discussion. Analysis of focus group discussions provided clarification of what respondents had intended in their questionnaire responses, gave some detailed explanations of why they had responded as they did, and of particular importance revealed information that had not been provided in the questionnaire responses. This information included statements from one female member of the group indicating that facilities for women, particularly toilets, were not being provided and from a male respondent that more thought should be given to safety, particularly how to avoid accidents while cycling and what should be provided if accidents did occur.

Case Study 10.2 discusses the two phases of mixed methods research, but what should be noted is that the second phase was not planned in detail in advance. The second phase, involving focus groups, was employed because it was felt that there was only limited data produced by the first phase, in terms of depth and detail of responses. Of particular importance also was that analysis of the results from the first phase had not provided a clear explanation of the reasons why respondents held the views that they had provided. Nevertheless, mixed methods can involve a context in which two phases of sequential research are planned in advance, and it is concurrent mixed methods research that can, and often does, involve the use of the two methodologies, that have been planned in advance. For example, Mason and Beamont-Kerridge (2003) when investigating attitudes to a music festival, initially used a questionnaire survey with visitors to the festival, but shortly after this research conducted a series of focus groups with local residents of the town where the festival was held. These two apparently separate pieces of research were part of planned concurrent exploratory research, although the respondents to the quantitative part were different from those respondents involved in the qualitative part. However, both groups were asked similar questions about their attitudes to the festival. In the analysis of the results of this research, a thematic approach was employed, using terms such as economic impacts, socio-cultural impacts and environmental impacts when referring to attitudes to the festival. This form of analysis was preferred to a discussion of results from one methodological approach, before a discussion of the results of the other approach. The major reasons for this were that discussing results from each methodological approach in turn, was considered to be too lengthy and would involve too much repetition, and that it would be easier to make comparisons between the results from the survey (quantitative methodology) and focus groups (qualitative methodology), in terms of similarities and differences, within a thematic approach.

Writing this chapter in your study

You will probably need a brief introduction reminding your readers why you used a mixed methods approach, which will provide the rationale for the use of mixed methods (and there is discussion of this in Chapter 6). This rationale will probably refer to the advantages and disadvantages of mixed methods.

The way in which you have conducted your analysis in terms of the sequencing of methodological approaches (or the use of a thematic approach) will need to be explained in relation to whether your use of mixed methods was sequential or concurrent. If a sequential approach was used, then the order of use (either qualitative or quantitative first) should be the guide for presenting your analysis. If you used a thematic approach to analysis, then this will need to be explained, with reference to why you used this approach and a discussion of the actual themes or topics used, including whether these emerged from the data or were ones you adopted and prepared in advance of conducting the research.

In relation to the analysis of quantitative data, you should refer to the discussion in Chapter 8 where there are details on analysing questionnaire surveys. In connection to the analysis of qualitative data, you should read Chapter 9 which is concerned with the analysis of data gathered from interviews and focus groups and discusses the process of constant comparison (Glaser and Strauss, 1967).

Any issues, including ethical issues, you had while conducting the research will need to be presented, as well as how you overcame any problems.

You will also need to provide details on the specific analytical techniques you used and the rationale for whichever ones you employed.

Student activities

1 If you made use of sequential mixed methods research, what would probably be the best way to present your analysis?

2 If you made use of concurrent mixed methods research, what would probably be the best way to present your analysis?

3 Why is using a thematic approach to data analysis often preferable in mixed methods analysis?

4 In relation to Case Study 10.1, create a table with two columns, the left column headed 'advantages of mixed methods research' and the right column headed the 'disadvantages of mixed methods research'.

5 What does Case Study 10.2 suggest about a) the advantages and b) the disadvantages of mixed methods research?

11 Conclusions

Introduction

The conclusions section of any academic publication is largely to provide a summary of what has gone before. So, the Conclusions chapter of your study is there to provide a summary of your previous chapters. In particular, it is meant to summarise your primary research in terms of the results, and analysis of these results. However, it is also meant to provide a summary of the key literature to which you have referred. Your primary research findings are intended to be summarised in relation to these key pieces of literature. In this way you can indicate where you have 'filled the gap' in the literature and made your 'contribution to knowledge'.

Contents of the Conclusions chapter in your study

The Conclusions chapter is intended to provide a summary of your findings, but also do more than this. The summary certainly needs to be there, but although it is a summary, it does not have to be mere repetition of what has gone before. This would be very boring reading and, more importantly, a missed opportunity to inform your reader of the importance of your findings. One way to do this is to

- Restate your aims and objectives near the beginning of the chapter,
- Indicate the key pieces of literature that influenced you in your primary research,
- State briefly how you went about conducting your research, in terms of methodology and specific techniques, and
- Allow this to lead into the summary of the main findings of your primary research.

Using the approach suggested above will enable you not only to summarise key points but demonstrate to your readers that you know that it is important to provide these statements in the final chapter of your study.

However, a key difference between what you write in this final chapter and what you have included in the 'Results' or 'Analysis of Results' chapters is that you will have had time to reflect on your findings and particularly the relationship between your aims and objectives, methodology and techniques, primary research results and analysis and the literature that contributed to your study topic. This reflection should enable you to discern the more important and the less important findings. It should also enable you:

- To see the advantages and disadvantages of your chosen methodology and techniques;
- To reflect on whether you would modify your overall approach and techniques, if you were to do the research again;
- To consider whether it is important or necessary to make recommendations based on your research.

The Conclusions chapter should have other sections, in addition to the summary. An important section should deal with the limitations of your research. Inevitably, your primary research will have been small scale, compared with most published studies. Although you may be under the impression that sending out 250 questionnaires, processing and analysing the 110 that were returned, was a major task for you. However, it is most likely that your sample, by definition, will have been small and hence your findings will not even be at the scale and significance of most published articles in academic journals. Therefore, it is not wise to over-claim for the importance of your results. A section of this chapter in which you outline the limitations is important, as you will be able to communicate to your supervisor and examiner that you are aware that you need to include reference to limitations, as well as actually indicating what the limitations are.

One other section that appears in the Conclusions chapter of many student research projects is titled: 'Recommendations'. This section is meant to include reference to possible future actions. These should be based on the research you have conducted and not 'guesswork' or thoughts you had, perhaps even prior to conducting the primary research. So, the evidence from your research should be used to create these recommendations. However, recommendations may not be necessary – it will depend on your aims and objectives and results. Perhaps the only recommendations that may be required are that you recommend that the research should be done again, for example at a different time, in a different location or with a bigger sample, or possibly, if you were to do the research again, you would change it in some specific way(s).

How to write this chapter in your study

Unfortunately, for too many students, the final chapter appears to be little more than an afterthought. It is a few lines in which some of the important findings may be repeated and a few recommendations (only some of which may be based on the actual primary research in the study) are made. When this occurs, it is a very sad state of affairs and is not likely to lead to a high assessment mark!

An understanding of human nature should inform you, that, as it is the last chapter of your work, you need to take this final opportunity to sell yourself to the reader. You need to maintain the reader's interest and leave them wanting to know more about your research, not because they are confused, but because they are interested and want to praise you for what you have produced. Hopefully, you will have done this in earlier sections of the study as well, but there will not be another opportunity to do this, after the final chapter.

As this is likely to be the last chapter that your examiner will read, he/she does not want to experience an anti-climax. You need to convince the reader that you have carried out a piece of research that has been conducted well and has added something (however small) to our understanding of a particular topic, theme or issue. Note that I have placed 'conducted well' before referring to 'adding something to our understanding' in the previous sentence. Remember, examiners are quite likely to be more interested in whether you conducted your research in a valid manner and achieved reliable results, than the actual details of what you have revealed. As I have written earlier, examiners are generally more likely to be experts in methodology, than experts in your topic.

■ The summary

In terms of what you need to include in this final chapter, you should be aware that a summary is certainly required. However, you do not need the detail provided in earlier chapters, but it should be a summary of the whole study. By this, I mean that you should include discussion of the key pieces of literature that you made use of, as well as summarising your own primary research. Too many students appear to be under the impression that they only need to summarise their primary research. The reason for the inclusion of a summary of the key literature, is, partly, that you can once again demonstrate your understanding of the literature and how it relates to your topic or issue. But, more importantly, by discussing in summary form this literature again, you can show where your primary research fits within the literature. Hence, you can indicate here how you have 'filled the gap in the literature' and made your contribution (however small) to our understanding of this topic, or issue.

You should not add in new findings to the Conclusions chapter. As a summary of the key elements of the study, there should not be the need for new material. However, it is possible that between starting your study and writing it up, a new article has been published on your topic, or issue. You can incorporate reference to this in the Conclusions chapter, but only if it is relevant to your study focus. Also make sure you indicate that it is a very new source of material that has been published since you began to conduct your research. If you do this, it should impress your examiner, as it is clear you are still reading material on your topic/issues, even after conducting your primary research.

However, do not write something to this effect: 'Having conducted my primary research and written it all up, I discovered last Thursday, by chance, an article in the major journal of...... published in 2005, in which the results are very similar to my research.....'. Your supervisor and probably your examiner will not be happy with this. They will argue that it is your responsibility as researcher, to seek out any appropriate material, particularly if it is in a major journal relevant to your topic, or issue, and you should have found this early on in your literature survey. However, if something like this has happened – you have recently found something of importance that was published a few years ago – you should try to incorporate reference to, and discussion of the article, in your Literature Review chapter.

In the summary that you produce, you should indicate only the major findings from your study. This means that you should summarise the key literature that you have used and its major points of relevance to your primary research. You should provide the key findings from your primary research. These findings should be linked to the key literature. Part of the skill that you are expected to demonstrate, in this final chapter, is your understanding of the key literature and how your primary research findings relate to this. In other words, there is no set guidance on the specific content of this part of the Chapter. You need to select what you regard as the key material and present it in a convincing manner.

Limitations of the research

Following on from the summary, an important element of this final chapter should be an indication of the limitations of your research. Unfortunately, many students fail to include any reference to limitations at all or provide little on the topic of limitations. If you do this, you will leave yourself open to the criticism that you have over-claimed for your findings. In terms of limitations, if you have used a questionnaire survey, remember to include some form of statement that indicates you are aware that it 'is a snapshot in time and space'. Even if you have largely replicated someone else's research, yours will have been different, particularly in terms of where it took place, when it happened and the size of the sample. You need to demonstrate that you are aware of this. If you have used a qualitative approach, you will hopefully be able to argue that you have achieved depth and detail in your responses. However, in terms of limitations, the number of respondents involved is likely to have been small, particularly in comparison with any funded projects on the topic and probably much smaller than the work of the major researchers in the field of your topic/issue. So, as in my comments about questionnaire surveys, do not over-claim for what has been revealed through your research.

If you are writing a PhD thesis or a Masters by research, you are probably required to have an oral component (usually known as a viva) to your examination. One of the favourite questions is: 'If you could do the research again, what changes would you make?' Even if you do not have a required oral component, thinking about what you would change, if you repeated the research, is a very good way of revealing the limitations of what you have done. However, also be aware when indicating your limitations, that you should not be so negative that you are over-critical of your research process and results. You should remember that it is important to convince your examiner that, although an apprentice researcher, you have produced generally well conducted research with interesting, valid and reliable results.

So, in summary, when you consider the term limitations, you may interpret this is meaning largely: 'What are the limitations of what my research has revealed?' However, in terms of your study, it is more appropriate to think of this as meaning; 'What are the limitations of my research, in terms of the overall research approach and techniques used?'

Recommendations

Many students regard it as important to make recommendations at the end of their study. On occasions, these far outweigh the conclusions. However, a key criticism of a number of students research that I have read and assessed, is that the recommendations are not based on the findings of the research. This situation can occur if a student has replicated another's study and uses the recommendations (inappropriately) from this. On other occasions, students appear to have made use of official reports, or other publications, which are on a similar topic to their study and inserted the recommendations from these sources into their Conclusions chapter – clearly these are not based on the student's own research. Additionally, some students seem to have felt that they should include recommendations and appear to have 'made something up' that could be regarded as relevant, even though it has not emerged from their research!

However, an important point to remember is that it is not essential to have recommendations in your study. If you have asked respondents for their recommendations in relation to your topic, then it is appropriate to include a summary of these in the Conclusions chapter (although you will also have included reference to these in your 'Results' and 'Analysis of Results' chapters). Some student research will have focused on topics that have recommendations 'naturally' following on from the conclusions, while in other studies, this may not be the case. The key message is, do not include recommendations if they are not based on your research findings. Nevertheless, one recommendation that you can make, if relevant, is that you would modify some aspect of the way you conducted the research. This could be in reference to particular research techniques, or the size or nature of the sample, or the timing and location of the research. It could also be useful to suggest that, for example, you could, in future, follow up the questionnaire survey you used, with a number of interviews. Be aware, it is not essential that you state that your research approach needs to be modified. So, you should first reflect on your results and overall findings, before deciding to include reference to modifying some aspect of your approach.

The last paragraph

At the very end of the Conclusions chapter, it is a very good idea to make a strong statement. This can be in the very last paragraph and can indicate something to the effect that despite the limitations of the primary research (you will have referred to this earlier in the chapter), your research has 'revealed something new', or perhaps it has 'confirmed something previously known via similar research, but in a new context' or 'with a different sample'. It could be that you are arguing that because you used a different technique to other studies, or a different methodology such as mixed methods compared with previous studies, you have revealed something new. Whatever you put in this last paragraph should be positive and leave the reader (your examiner) believing that they have read a worthwhile, valid, reliable and (hopefully) interesting piece of research. In other words: 'Go out with a bang, rather than a whimper!'

Student activities

1 'The Conclusions chapter is mainly an opportunity to restate your major primary research findings'. Indicate whether you agree or disagree with this statement, providing reasons for your response.
2 Why should you not introduce new findings in your Conclusions chapter?
3 Why is it necessary to include the limitations of your research?
4 Is it appropriate to refer to your literature review in the Conclusions chapter? Give reasons for your answer.
5 The following phrase has been written above: a 'final opportunity to sell yourself to the reader'. What do you understand by this phrase?

References

Agarwhal, S (1997) The Resort Life Cycle and Seaside Tourism, *Tourism Management*, **18** (2) 65-73

Berg, L (1967) *Acquiescence and Context in Response Set and Personality Assessment*, New York, Aldine Publishing

Bernard H.R. (2000), *Social Research Methods: Qualitative and Quantitative Approaches*, Sage, Thousand Oaks, London.

Brotherton, B (2008) *Researching Hospitality and Tourism: a student guide*, London, Sage

Butler, R (1980) The Concept of a Tourism Area Life Cycle, *Canadian Geographer* ***24** 5-12*

Clarke, M. Riley, M Wilkie, E and Wood, C (1998*) Researching and Writing Dissertations in Hospitality and Tourism*, London, Thomson

Cohen E. (1972), Toward a sociology of international tourism, *Social Research,* 39, 164-182.

Cohen, L and Manion, L (1995) *Research Methods in Education,* London, Routledge

Coles, T Duval, D and Shaw, G (2013) *Writing Dissertations and Theses in Tourism Studies and Related Disciplines*, London, Routledge

Cooper D and Schindler P (1998) *Business Research Methods* (6th Ed) Singapore, Irwin/McGraw Hill

Couch, A and Heniston C (1960) 'Yeasayers and Naysayers: agreeing response set as a personality variable', *Journal of Abnormal Social Psychology*, **60** 151-174

Creswell, J (2009) (3rd ed.) *Research Design: Qualitative, Quantitative and Mixed Methods Approaches*, London, Sage

Creswell J.W. (2018), *Research Design: Qualitative and Quantitative and Mixed Methods Approaches*, (5th ed.), SAGE, Thousand Oaks, California.

Creswell J.W. and Plano Clark V., (2003) *Designing and Conducting Mixed Methods Research*, SAGE, Thousand Oaks, CA

Creswell J.W. and Plano Clark V., (2007) *Designing and Conducting Mixed Methods Research*, 2nd Ed SAGE, Thousand Oaks, CA

Dann G. (1977), Anomie, ego–enhancement and tourism, *Annals of Tourism Research*, 4, 184–194.

Davies B. (2003), The role of quantitative and qualitative research in industrial studies of tourism, *International Journal of Tourism Research*, 5, 97–111.

Decrop A. (1999), Triangulation in qualitative tourism research, *Tourism Management*, 20, 157–161.

Eldabi T., Irani Z., Paul R.J., Love, P.E. (2002), Quantitative and qualitative decision-making methods in simulation modelling, *Management Decision*, 40, 64–73.

Evans, J and Mathur, A (2005) The Value of On-Line Surveys, *Internet Research*, April, 195-219

Garrod, B and Fyall, A (Eds) (2013) *Contemporary Cases in Heritage Tourism: Volume 1* Oxford, Goodfellow Publishers

Getz, D (1986) Tourism and population change in the Badenoch and Strathspey district of the Scottish Highlands Scottish, *Geographical Magazine*, 102 (113-126)

Getz D (1994) Resident attitudes towards tourism: a longitudinal study of the Spey Valley, Scotland, *Tourism Management*, 15 (4) 247-258

Glaser, B and Strauss, A. (1967) *The Discovery of Grounded Theory: Strategies for Qualitative Research*, Chicago, Aldine Publishers

Guba E. (1987) What have we learned about naturalistic evaluation? *Evaluation Practice*, Vol. 8, pp. 23-43.

Jennings G. (2011), *Tourism Research*, (2nd ed.), John Wiley & Sons, Brisbane.

Hall, C.M. (2002) Travel Safety, Terrorism and the Media: The Significance of the Issue-Attention Cycle, *Current Issues in Tourism* **5** (5) 458-466

Hammersley, M (1981) Martyn Hammersley Files Teaching qualitative methodology, finalfff.doc (wordpress.com)

Hewitt-Taylor, J (2001) Use of constant comparative analysis in qualitative research, *Nursing Standards* 42 39-42

Khoo-Lattimore K., Mura P. and Yung R. (2019), The time has come: a systematic literature review of mixed methods research in tourism, *Current Issues in Tourism*, 22, 1531-1550.

King, G., Keohane, R and Verba, S (1994) *Designing Social Inquiry*, Princeton New Jersey, Princeton University Press

Krueger, P (1994) *Focus Groups*, London, Sage

Krueger, P (2004) *Focus Groups*, (2nd ed) London, Sage

Kuhn, T (1970) *The Structure of Scientific Revolutions* (2nd ed.) Chicago, University of Chicago Press

Lincoln N. and Guba, E. (1985), *Naturalistic Enquiry*, Sage, Newbury Park.

Long, J (2007) *Researching Leisure, Sport and Tourism: The Essential Guide*, London, Sage

Mason P (2001) *Wellington Zoo: Visitor Survey and Keeper Evaluation; A report to the Zoo Management Committee*, New Zealand, Dept. of Management Systems, Massey University

Mason P (2003) *Tourism, Impacts, Planning and Management*, Oxford, Butterworth Heinemann

Mason P (2008) *Tourism, Impacts, Planning and Management*, (2nd ed) Oxford, Elsevier

Mason P (2014) *Researching Tourism, Leisure and Hospitality for your Dissertation*, Goodfellows, Oxford,

Mason P. (2017), *Geography of Tourism: Image, Impacts and Issues*, Goodfellow, Oxford.

Mason P and Beaumont-Kerridge, J (2003) Motivations for Attendance at the 2001 Sidmouth International Festival, in Long, P and Robinson M. (Eds) *Festivals and Tourism: Marketing, Management and Evaluation*, 33-46, Sunderland, Business Education Publishers.

Mason P and Cheyne, J (2000) Resident Attitudes to a Tourism Development, *Annals of Tourism Research* **27** (2) 391-411

Mason P and Kuo I, (2007) Stonehenge: International Icon or National Disgrace? *Journal of Heritage Tourism* **2** (3) 168-183

Mason P, Grabowski, P and Wei D (2005) SARS, Tourism and the Media, *International Journal of Tourism Research* **7** (1) 11-22

Mason P, Augustyn A and Seakhoa-King, A (2010) 'Exploratory Research: The first stage of sequential mixed methods research' *International Journal of Tourism Research.* **12** (5) 432-448

Mason, P, Augustyn, M and Seakhoa-King, A (2021) Mixed Methods Research in Tourism; a Systematic Sequential Approach *Folia Turistica*, 53, 9-31

Maykut P. and Morehouse R. (1994), *Beginning Qualitative Research: A Philosophic and Practical Guide*, The Falmer Press, London.

Moser, C and Kalton, G (1989) *Survey Methods in Social Investigation*, Aldershot, Gower Publishing.

Neumann, W (1984) *Social Research Methods: Quantitative and Qualitative Approaches*, Needham Heights, Allyn and Bacon.

Parks, T (2003) *A Season with Verona*, London, Random House

Patton (1990) *Qualitative Research and Evaluation Methods*, (2nd ed.) Thousand Oaks, California, Sage

Patton (2002) *Qualitative Research and Evaluation Methods*, (3rd ed.) Thousand Oaks, California, Sage

Phillimore J. and Goodson L. (2004), Progress in qualitative research in tourism: epistemology, ontology and methodology. [in:] Phillimore J., Goodson, L., eds., *Qualitative Research in Tourism: Epistemology, Ontology, and Methodologies*, Routledge, London, pp. 4–29.

Raybould, M Digance, J and McCullough C (1999) Fire and Festival: Authenticity and Visitor Motivation at an Australian Folk Festival, *Pacific Tourism Review*, **3** 201-212.

Riley M (1996) *Attitudes and Attitude Measurement*, London, Taylor and Francis

Riley R.W and Love L.L. (2000), The state of qualitative tourism research. *Annals of Tourism Research*, 27, 164–187.

Ryan C (1995) *Researching Tourist Satisfaction*, London Routledge

Sapsford, R and Jupp, K (Eds.) (1996) *Data Collection and Analysis*, London, Sage

Seakhoa, A., Augustyn, M. and Mason P (2021) *Tourism Destination Quality: Attributes and Dimensions*, London, Emerald

Smith J, and Heshusius L. (1987), Closing down the conversation, The end of the quantitative-qualitative debate among educational researchers, *Educational Researcher*, 15, 4-12.

Stake, R (1995) *The Art of Case Study Research*, Thousand Oaks, California, Sage

Taylor, J and Edgar D (1999) Hospitality Research: The Emperor's New Clothes. *International Journal of Hospitality Management* **15** (3) 211-227

Tashakkori, A and Creswell, J (2007) *Handbook of Mixed Methods Research in Social and Behavioural Research*, London Sage

Tashakkori, A and Teddlie. C (2003) (Eds) *Handbook of Mixed Methods in Social and Behavioural Research*, Thousand Oaks, California, Sage

Teddlie, C and Tashakkorie, A (2009) *Foundations of Mixed Methods Research*, London, Sage

Thomas, J and Nelson, J (1990) *Research Methods in Physical Activity*, (2nd Ed) Champaign, Illinois, Human Kinetics

Timans R., Wouters P. and Heilbron J. (2019), Mixed Methods Research: What is it and What could it be?, *Theory and Society*, 8, pp. 193-226.

Veal T (2011) *Research Methods for Leisure and Tourism: a Practical Guide*, (4th ed) Harlow, Prentice Hall

Walle A.H. (1997), Quantitative versus qualitative tourism research, *Annals of Tourism Research*, 24, 524–536.

Watson, G (1987) *Writing a Thesis*, London, Longman.

Weber, R (1990) *Basic Content Analysis*, London, Sage.

Whyte, W (1982) 'Interviewing in Field Research' in Burgess, R (ed.) *Field Research: A Sourcebook and Field Manual*, London, Allen and Unwin, 111-122.

Whyte, W (1984) *Learning from the Field*, Newbury Park, California, Sage

Yin, R (1994) *Case Study Research: Design and Methods*, (2nd Ed.) Thousand Oaks, California, Sage

Index

www.ingramcontent.com/pod-product-compliance
Ingram Content Group UK Ltd.
Pitfield, Milton Keynes, MK11 3LW, UK
UKHW051847100926
8556IPUK00005B/505

9 781915 097002